T0356702

WHEN
THE EARTH
WAS GREEN

Also by Riley Black

FOR ADULTS

The Last Days of the Dinosaurs

Written in Stone

My Beloved Brontosaurus

Skeleton Keys

Deep Time

FOR CHILDREN

Prehistoric Predators

Did You See That Dinosaur?

WHEN
THE EARTH
WAS GREEN

Plants, Animals, and Evolution's
Greatest Romance

RILEY BLACK

ST. MARTIN'S
PRESS
NEW YORK

First published in the United States by St. Martin's Press,
an imprint of St. Martin's Publishing Group

www.stmartins.com

Designed by Steven Seighman

Illustrations by Kory Bing

The Library of Congress Cataloging-in-Publication Data is available
upon request.

ISBN 978-1-250-28899-8 (hardcover)
ISBN 978-1-250-28900-1 (ebook)

First Edition: 2025

10   9   8   7   6   5   4   3   2

For Jet
On the outcrop or on the couch, you are such a good boy.

Contents

The tree that moves some to tears of joy is to others a green thing that stands in the way.

—WILLIAM BLAKE

GEOLOGIC TIMELINE

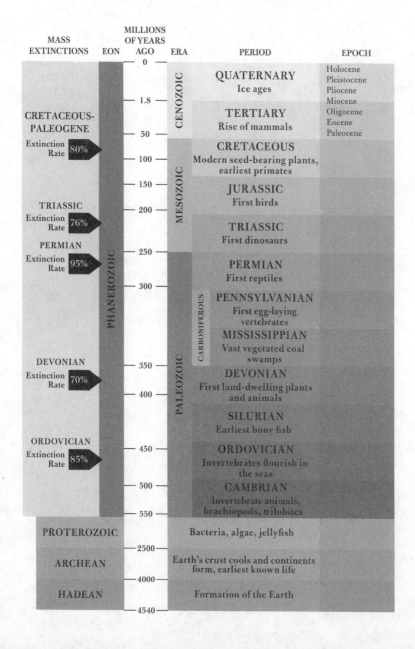

Introduction

I WAS SUPPOSED TO BE LOOKING FOR *T. REX*.

You'd think that the petrified bones of a flesh-rending reptile the size of a school bus wouldn't be all that difficult to find. I imagined curved ribs and smiling jaws jutting out of a stony and weathered butte as if the immense saurian had been waiting 66 million years for its moment in the sun, ready to be coaxed from the encasing stone. But searching for a dinosaur is anything but certain. A previous expedition by the same museum crew had turned up a smattering of toe bones and fossilized shards, enough to know that the "king" was in the building, but little else. Our search was left at the mercy of what erosion had wrought ever since those ancient strata were exposed by the rise of the Rocky Mountains. Perhaps there was more in the rock. Perhaps those little piggies were all that was left.

Fossil hunting will test the limits of your optimism. Every morning you wake up within the gently fluttering walls of your tent, the air already too hot at 8:00 A.M. as you scratch the scabs that were fresh bugbites yesterday afternoon. As you fill your water bags and stuff snacks into your pack, you make peace with the fact that your narrow field of human senses are probably going to miss the critical cue that will give away the prehistoric remains you've traveled so far to find.

Boots on the rock, you take one step after the next and try to remind yourself that, if you're lucky, you might literally trip over the best find of your life. Spectacular skeletons have been unearthed when someone spotted little more than a tail-bone peeking out of the dirt, and more than one expert has washed away the rime of obscuring stone by taking a piss. If only you might be so lucky. There are no game trails to follow or recent scat from which a prehistoric organism's whereabouts might be drawn out. The ancient remains hold still, preserved and confined in the stone, somehow harder to find than a living animal that can flap, hop, run, crawl, or swim away from you. All you can do is learn to love the search, using the fact that no one else seems to be finding much, either, as an emotional salve.

I knew all this, but still I wanted to hope. I told myself that I'd even settle for a broken and isolated tooth, what remained of a serrated, biological ice pick the size of a banana. An adult *Tyrannosaurus* had about sixty teeth in their mouth at any one time, after all, and the fearsome reptiles automatically replaced their old teeth with fresh choppers throughout their lives. Finding one that had slipped from its socket to be buried in the gray hillside didn't seem like an unreasonable request. Under the Montana sky, a dark curtain of clouds gradually burned away to reveal an endless, inverted sea of blue. I walked, poked at potential fossils, turned over rocks, and hoped to spot some unusual color or texture out of the corner of my eye that would signal the rock before me wasn't what it seemed. Over time, however, I had begun to feel like I was most skilled at finding where the tyrannosaur was *not*, narrowing the field of possibilities for where a four-foot femur or a dentary dotted with teeth might be hiding. And the team had other places to visit. Looking for dinosaurs on public lands requires specially arranged permits, patches of rock sometimes hours away from each other across the parched splay of western desert. Time was running out.

When you're looking at a slope where a dinosaur might be tucked snug into the stone, the most logical thing to do is start at the bottom. Walk slowly and carefully along the base of the incline, paying special attention to any washes and arroyos for isolated bones or even fossil fragments that might give away the presence of something better. If you're fortunate, you can follow the trail of petrified bread crumbs up to their source. And if the sun is at the right angle, the Earth is the correct distance from Mars, the wind is just at the proper speed, or whatever it is that determines luck, the osteological hash will lead you to a lovely leg bone jutting out of a cliff or a string of barely eroded vertebrae still in place.

I was not lucky. Textbook procedure had entirely let me down, and so I decided to break the rules and find a way up the slope to a shelf of resilient stone the color of dried blood jutting out at the very top of the section. Fossils found in such rock are rarely pretty, sometimes only coming down as impressions of bones that have eroded away rather than the animal's remains themselves, but that would be better than nothing. I clambered to the lip of the ancient stone above the slope, considering whether the sky really did seem bigger in Montana than elsewhere, as I pried the sharp end of my rock hammer into the stone to lever up a stony sliver in frustration. "Come on, girl, where are you hiding?" I murmured to myself.

When the sheet of Cretaceous sandstone popped and flipped, there was a leaf just below it.

For a moment, I thought I had made a mistake. The leaf looked almost new, as if it had just recently wafted down on a late summer breeze and slipped under the sedimentary layers. I could see everything down to the tiny, branching veins that spread toward the leaf's multiple points. It didn't look all that different from the maple leaves that collected along the curb every autumn of my East Coast childhood. How could this be here? Was there some gap in the stone that allowed last year's deciduous leaves to sneak underneath the layers and become compacted in a cushion of organic debris? The spreading tributaries of the plant's veins looked too delicate to be fossil. But this wasn't some soppy autumn leaf I could easily pull off the rock, and there was no whiff of organic decay. Not to mention that among the surrounding landscape nothing stood taller than waist-high sagebrush. There were no trees here to shake off such leaves. The piece of organic debris was literally set in stone, a deep and pervasive shade of rust instead of the vibrant green that had been touched by Cretaceous sunlight. Maybe I had missed the dinosaur, but I couldn't hold back a toothy grin of my own. This was better.

Wandering out into the desert and searching for a specific dinosaur species is a fool's errand. Paleontology has survived as a discipline because experts have learned to be grateful for what the fortunes of preservation and the luck of discovery have brought to them, a field built upon what we just happen to find. It's not all that different from when you go to the park as a kid and come home with a curiously smooth pebble or a katydid in a jar that you hadn't known existed until you happened upon it. And even then, what are such treasures outside of their setting? You can learn a katydid's shape through the glass of a mason jar but not how the animal lives; just as you can appreciate that the pebble came from the riverbank but not the motion of the water that polished it down into something to keep on your bookshelf or the billions of years of geologic history that led to its formation. The same holds for dinosaurs and myriad other forms of prehistoric life. A *T. rex* that's excavated, plastered, hauled away from its home rock, driven to a museum, unpacked, and carefully freed from its ancient sediment tells you about the animal's skeletal system and perhaps a smattering of other biological cues, but precious little about the world that animal inhabited—all the ways the life of that one, unique *Tyrannosaurus* intersected with the Earth of 66 million years ago.

Alone, a dinosaur is meaningless. We put them on literal pedestals in our museums, behind glass and railings as befits their place as the world's most ancient and long-standing celebrities. But what is a *Tyrannosaurus* without a forest to conceal its shadow as it stalks? What is a *Triceratops* without a buffet of ferns and cycad fronds to eat? What is a Mesozoic world without the busy machinations of pollen-collecting beetles and nectar-drinking butterflies that assisted the innumerable, vegetal lives that set the basis for so many other living things to exist? The fossil leaves I turned up from those tan rocks in Montana said more about the world of *T. rex* and her neighbors—the temperature, the rainfall, and forest—than a multi-ton enigma of which we are still striving to understand the biological basics. Even if you had complete skeletons of every single dinosaur species in a specific environment, imagining how they interacted would be a glorified version of mashing together dinosaur models in a sandbox. We need to know the stories of those smaller, but no-less-vibrant lives that surrounded and fed the terrible lizards. Ripped away from Earth's botanical history, life on our planet doesn't make the least bit of sense.

What's true of the past also applies to the present. We often depict plants as little more than the static background for animal behavior—grasslands are sustenance and setting for gazelle, rafts of duckweed conceal lurking alligators, and tigers need forests of the night to burn so brightly in. Even within the confines of our own experience, plants are often part of an inert-seeming landscape until they have some direct effect on our day-to-day lives. We'll get annoyed at annual influxes of pollen that coat our cars and send us into sneezing fits, weeds growing in the yard embody a hated chore to get around to some mild Sunday morning, and vegetables become dietary homework shoved to the sides of our plates, even as we close our eyes in relief under the shade of a summertime tree and drive out of our way to catch the one autumn day when every stand of forest seems aflame with color. We recognize that plants are alive, as we are, and yet we hold them apart. Plants become lives without mind and feeling, not so different from the soil out of which they grow. Who would stop to ponder the life of a Woods' rose or even a blade of grass? We are surrounded by all this verdant life—our own existence dependent upon it—but we generally lack the patience to confront what we don't know about the maple growing through the fence or the sunflowers raising their fibrous stalks from a roadside ditch. Our ignorance certainly colors what we understand of life's ever-expanding evolutionary panorama. Palms and conifers might frame a prehistoric scene where bizarre reptiles snarl at each other, but we tend not to ask about the lives that make up the habitats in which our favorite prehistoric creatures roamed. Such scenes are absolutely bursting with life, but it's challenging for us to think beyond the experiences of the toothy and reptilian.

Take a step back for a moment. Imagine your favorite dinosaur—or some other terrestrial creature if the saurians are not to your personal taste. Think of that animal carefully stepping through the evening forest, the low-angled glow of the sun sending shafts of orange and gold through the shadowy rows of tree trunks. That animal is moving through a grove of *living things*, not just knowable as individual plants but lives that are literally intertwined from canopy to runs sunken into the ground. Leaf-munching insects, lichen dotting the bark, fungus growing along the roots, and more, discernible from each other but all connected. The entire scene is alive, a tangle of stories so varied and intricate that you could start following a single, small thread of existence and spend the rest of your life tracing the snarled, pulsing web of life it's part of. As we imagine our prehistoric friend taking each step, we can envision that foot pressing into mud full of decayed plant matter, the whisper of the leaves as a breeze rustles the botanical neighborhood, the smell of flowers eager for pollinators—not simply envisioning the forest but sensing it. What I'm asking you is not just what you see in your mind's eye, but what an ancient forest might have *felt* like.

Scientific results can help us populate the scene. Ways to identify, compare, analyze, and catalog nature have helped introduce us to the varied and wonderful lives around us and how those organisms are unique in all of history. But the journey we're about to take together requires something more. Even though there is certainly so much more to learn, we have no shortage of papers and tomes trying to boil down and simplify millions upon millions of years of life's history into something we can understand, like the copy of *Common Fossil Plants of Western North America* sitting on my desk as I write this. But such information is like a smattering of spices and ingredients laid out on the countertop before you start cooking your favorite recipe—those individual parts, even arrayed side by side with each other, don't really feed the curiosity that drives the whole process to start with. Emotion and imagination are the heat that transforms all those parts into visions of the past that have kept people like me picking away at hot stone for generation after generation, not just wanting to see what's left of ancient worlds but desiring connection to times and environments we'll never witness ourselves. Each and every fossil is a touchstone, no matter whether it's a fragment of an ancient leaf or the complete skeleton of a whale, inviting us to consider lives that have come and gone long before us. We may even learn something about what might evolve after we're gone.

Dinosaurs and other charismatic megafauna have often been employed as such inspirations for thinking through time. Such creatures seem so strange and grand that we can't help but wonder about when they lived and why they're not here anymore. If I had found that elusive *T. rex* along the Montana hillside, the animal's remains would have embodied the fact that the Earth is very old, life has changed through time, and extinction is a reality. The consequences of such finds launched the entire field of paleontology to uncover and understand the nuances of these amazing truths. Even so, the bare facts of life's extremely long history leave us with many more questions. We want to know what these species were like when they were alive, questions about their day-to-day existence that requires placing them within their home habitats to fully imagine. Consider the tree leaves I peeled out of the stone from the tyrannosaur's last resting place, pressed and preserved by the Earth itself. I wonder how the living tyrannosaur would have perceived the trees those leaves belonged to. Did they have a scent, their fruit perhaps emanating a carrion stink to attract dinosaurs, bugs, and mammals to eat them? Did the leaves fall in the autumn? Did they shade great scaly backs in the height of summer? Did they lash back and forth as thunderstorms rolled across the swamps and glades of the Hell Creek ecosystem? Might they have helped the monstrous carnivore conceal itself within the shifting beams of scattered sunlight as it stalked? We know such scenes must have transpired so very long ago, everyday moments when life touched life.

Plants are often the missing part of our paleontological daydreams. For years, I've been concerned with the comings and goings of various prehistoric animals, especially the vast dinosaurian menagerie. Many of the questions paleontologists ask about these animals involve their relationships to each other—who is related to whom, the thrust and parry of attack and defense, and what species comprised the varied dinosaurs that have thrived from pole to pole for more than 235 million years. On expeditions, plant fossils I've stumbled across—in Montana, Utah, New Mexico, Alaska, and elsewhere—have been treated as little more than prehistoric debris, telling us not much beyond the fact that plants once grew here as they do over much of Earth's landmass. Only exceptional botanical fossils seem to merit a note in the field book or collection. The experts I've accompanied have primarily been interested in dinosaurs and other animals, a tree of life determined by the relationships between bones. And yet every time I wondered what the daily existence of those creatures must have been like, the moment would be inscrutable without plants. I started to see fossil plants as more than just filler in the ecological background. Understanding individual plants and habitats opens into broader vistas that span Deep Time, shifting relationships that make our present moment in time all the stranger. The world becomes wilder when you realize that what you perceive around you is the outcome of untold happenstances that flowed one into the next. Life as we know it did not have to exist, and yet it does, the legacy of mass extinctions and the continual unfolding of evolution forming the essential story of why you, I, and the mint bush growing in my front yard are all present in this thin slice of time.

The fossil record not only introduces us to the players and the general arc of the story, but reveals snippets of paleontological dialog where we can draw out interactions both fleeting and formative. Plants and animals have a drama all their own. Animals never would have crawled out of ancient bogs without scaly trees and other plants that altered the terrestrial realm first, thick and otherworldly forests where crunchy insects would eventually entice our fishy ancestors to belly flop onto shore. Gigantic dinosaurs would not have grown to such prodigious sizes without a surplus of green food. Insect life would not be quite so diverse or colorful without plants to pollinate. I had been looking at ancient ecosystems from the top down, following in the steps of charismatic megafauna. I had ignored the flowering of life beneath their feet and that enclosed their worlds—the understory.

The more I came to learn about the tale, the more I wanted to know about these players. Animals were relatively easy to empathize with. As different as I am from a spike-tailed *Stegosaurus* covered in bony plates, we share the same basic vertebrate setup. Even the most familiar plants, by contrast, are so different from us that it's challenging to imagine what universals such as sun, wind, and rain might mean to them. I can see that the leaves of the plants of my desktop terrarium grow toward the light, but what is that experience like? The further we get from our own evolutionary neighborhood, the harder it is for us to connect or understand all the different ways there are to be alive. Plants are the aliens that live in the yard. They grow according to timescales that are often imperceptible to us, release the most essential element in the air we breathe as a waste product, cast their sex cells to the wind, and communicate with each other in ways we've only just barely begun to perceive. The great trick of the plants is that they are so ubiquitous, so essential for our own lives, that we've ceased to be impressed by their lives and how ours intersect with theirs.

Life on Earth is so varied and strange that even if every biologist picked just one species to study for their entire lives, we'd still miss so much. That's because such a divide-and-conquer approach would only present us with a cast of characters without much insight into how all the players have created a story. Part of what makes life on our blue-green marble continue to change is interaction, or what we might think of as the domain of ecology. A rose growing in the garden is not merely a plant. It's a living thing that rests at the intersection of the unintentional gardening of weather, climate, the hunger of insects, and even the evolution of what we consider beautiful. Pick any living thing on this planet and you have grabbed hold of a thread that flows into an ever-shifting tapestry, beautiful close-up and far away. And because so much of our history—of life's history—has been dependent on plants, beginning with their story can open spectacular views into how the life we see around us came to be as it is. The story is not about origins, because every possible starting point would have something else beyond, but instead flows through moments of an evolutionary romance that continues to flower. We did not arrive here on our own, but as part of an ongoing relationship with the botanical.

Just as I've reminisced with partners about key moments in our shared history, I've written this book as a series of unfolding vignettes that speak to how animal and plant life have changed each other through the ages. Some moments changed the world to the point that we can identify the influence of plants through how they altered the composition of the air and even the nature of the planet's stone. Others are smaller stories that are nevertheless part of the evolving dynamic between animals and plants, such as when ancient catnip evolved anti-insect defenses that had an unexpected consequence for felines, or when carpets of fallen leaves began to provide refuge for overwintering invertebrates. Plants are even responsible for parts of the fossil record we otherwise wouldn't have, such as insects and other small creatures encased in amber. And as strange as some of the organisms or environments in the following pages might seem, all have a connection to the world we live in now. How tree trunks become petrified fossils, the way plant-munching herbivores convert plants to climate-nudging methane released into the air, and what life in the trees has to do with the form of our own bodies creates a reciprocal relationship in which the past informs the present and the present can help guide our questions about the past.

I've endeavored to restore these ancient scenes with as much accuracy as I can grab hold of and carry. Each chapter has its own appendix entry, detailing what we think we know, what we might guess at, and what simply struck my fancy. Use the information in any way or order you see fit, perhaps as seeds for your own curiosity. My hope is that the background information will act as an expanded version of scientific show-and-tell that underscores what we know, what we suspect, and what we've yet to find out. This is how I celebrate science, each new thing we've come to know prompting even more questions that keep us conversing with nature. And when I say that this is a tale of evolutionary romance, I truly do think of it in terms of relationships that transform each part of themselves as they interact. In place of the traditional paleontological narratives of domination, conquest, and colonization, I want to look at the same layers with community in mind. It's the interactions between living things that continually alter and refresh life's diversity. Consider pollination: Plants evolved flowers to better reproduce with each other, with animals acting as unintentional assistance as they wandered in search of pollen or nectar. Over time, some plants came to rely on particular animals to pollinate them, those animals in turn becoming dedicated pollinators, a novel niche that could not have existed before. Diversity continues to create more diversity in this way. The more living things there are, each unique and variable, the greater the chances that small happenstances can open new possibilities. This is how forests are made, and it's why life on Earth is not a monoculture. Even if everything on the planet were to be reduced to a primordial soup populated by a single species of bacteria, variation and chance would immediately start to cause those microorganisms to start developing new ways of living, new specializations, that would eventually create a whole new timeline for life on this planet. Life evolves, and the greater variety that evolution spins off, the more possibilities there are for life to take on new shapes. It's part of why the stories in these pages present a changing cast of characters. Evolution doesn't proceed along a predetermined course where we can expect woolly mammoths every hundred million years. When life begins to do something new, whether that's photosynthesizing or developing symbiotic bacterial relationships to digest plant material, an entire timeline of new possibilities opens.

What follows in these pages is my attempt to explore those twining threads and tendrils in life's ongoing story. I had to accept at the outset that this drama is flawed and fraught. I am, after all, only human. Our curiosity about other living things is limited by the array of senses we've been granted through our own evolutionary lineage. I can't see in ultraviolet colors. There are innumerable pheromones I cannot detect or respond to. Electromagnetic fields may as well be invisible magic as far as my body can actively perceive. If I had any of these abilities that are common to other animals, my insights might be different. That's hardly all. I'm a large mammal—technically megafauna. If I were the size of a bacterium, a mouse, or an elephant, my thoughts on plants and what they are would be shaped by my size and other aspects of my biology. Even considering some of my most direct interactions with plants during mealtimes, the plants I eat come from farms and have been cultivated over centuries to often be seedless, larger, or tastier than the original species. Such idiosyncrasies in experience and perception surely shape the story, knowledge that anyone fond of science eventually has to grapple with. We exist within a natural reality that we can question, investigate, and find inspiration in, but even what we know is not the whole of what exists.

It's taken time for me to learn and appreciate all these limitations, and more, affecting even the questions I might think to ask. Such constraints give form to our wonder, and that is what I wish to share with you. Within these pages, I hope you encounter living things you've never heard of before today as you slide from one window to the next, and each is an illustration of a broader story. Plants act as our focal points, mainly because they make such beautiful examples for considering how they've changed the world and been shaped by it in turn. The root system beneath each tale is the deeper story of how the relationship between plants and the rest of our planet has unfolded, the constant process of making and unmaking that unfolds as the green shoot of life continues to stretch from its point of origin. What I mean by "when the Earth was green" is not some irretrievable past or a denial that, despite our efforts, there is a great deal of green around us today, but thinking of key moments that plants have changed the nature of nature itself just as we tend to pay special attention to the great springtime blooms when the first leaves unfurled all around us are so impossibly, vibrantly verdant that I can't help but smile when I notice the hills around my home burst with color. When I see those shades, I think of all those varied lives and how many different ways of existing there are. If we take a moment to find connection with and live through lives unlike our own, we can find worlds within worlds just like rows of petals in a flower. Happy accidents, chance encounters, and nature's persistent habit of growing outside the boxes we try to organize it within have created relationships of incredible, transcendent change, moments in a broader tale that will keep unfolding for as long as there is an Earth. A bloom is the happiest of accidents.

1

Sex in the Shallows

1.2 billion years ago
Arctic Canada

LITTLE RED THREADS WAFT BACK AND FORTH TO THE slow rhythm of the waves. Each is so incredibly small that every wispy strand is anchored between individual pieces of crushed stone and quartz, relying upon what can only be a tenuous hold on the ever-shifting sea bottom. They can do little else but dance to the time called by the sea.

To pick out any single filament from the mass would be extremely difficult. These living things, stacks of cells that would seem to barely reach above the sediment, only span a couple of microns each, gathered around each other in organic strings growing from the tiny spaces between sand grains that were once stone. They make their own food by using energy from sunlight to rearrange organic molecules into forms able to nourish them during their sedentary existence. Despite their humble appearance, however, the threads are something novel on Earth. In a world of singular, self-contained bodies, the algae's cooperative cells can carry out different functions—whether that's holding tight to the substrate or sending a new generation out into the world. Their messy clumps are a foreshadowing, in miniature, of forests that will stretch both into the air and deep below the ground.

The strands are not Earth's first photosynthesizers, nor are they technically plants. The multicellular strings are something between, a new expression of forms that have long existed but arranged in such a way that each part of the organism has its own role to play. The filaments have been shaped by the shifting demands of natural selection as well as a surprise cohabitation, a history in which the barriers between one living thing and another have become so permeable as to almost be intimate. And for the moment, the meek little algae are among the most complex life-forms among what is still—and will always be—a world densely populated by microbes.

So far as sheer longevity is concerned, organisms like these red algae will persist across incredible spans of time. *Bangiomorpha*, as they will eventually be named, wouldn't look out of place in a twenty-first-century tide pool. But all those other organisms that we might associate with a trip to the beach are still a long, long way off. During this Mesoproterozoic time, there are no shells in the sand. Nothing large enough to be visible to the human eye swims through the water. No fish, no cephalopods, no bivalves, nothing at all that we would recognize as animal lives in these waters—or anywhere else on Earth. The earliest recognizable animal life, sponges that will form their own anchors to the ocean bottom, are still over 400 million years off in the future from this point in prehistoric time. And above the waterline, sand and stone gives way to a landscape of bare rock beyond. Snow-capped mountains in the distance are practically bald; no trees or other vegetation cover their slopes. Multicellular life hasn't gotten there yet. In all, a visitor to this time might consider the planet lifeless despite all the tiny organisms constantly reproducing themselves, eating each other, and taking in the sun. Earth is an ocean planet, cradling most every living thing within the salty womb of the sea.

By this time, with *Bangiomorpha* growing in the shallows, it's been more than two billion years since the earliest cells coalesced. During this incredible span of time the descendants of that first life have been busy reproducing, evolving, and filling the world's waters with each successive generation. All the changes among these living things—from how they gather food to what conditions are most suitable for them to thrive—are setting up the shape of life in the hundreds of millions of years to come afterward. The possible array of forms life on Earth might take is already being influenced by the details of these microscopic lives. Consider it this way: The time between the first life on Earth and something as basic as *Bangiomorpha* is more than 2 billion years, while the amount of time between *Bangiomorpha* and the earliest humans is about 1,194,000,000 years. Most of Earth's history is told through the lives of individual cells. It's only in this moment that what were once single cells are beginning

to combine and coalesce into new and unexpected arrangements, the essential foundations for everything that will
follow.

On an overcast day like this one, skies weighted with a
puffy gray blanket of clouds in every direction, the small
strings simply shimmy in place and wait for the life-giving
sun to warm the shallows again. They can't make food for
themselves without fresh photons to jump-start the essential
process. But a human perspective is inadequate to understand
life during this incredibly ancient slice of Earth's history. Life
is ubiquitous during this time, what will be called the Stenian Period by geologists, but it thrives at a scale that would
require microscopes to see, much less understand. For more
than two billion years, since something *alive* originated on
the planet, biodiversity has existed almost entirely as arrays
of single cells replicating themselves *ad infinitum*. Mats of
bacteria and aggregations of cyanobacteria striving for sunlight are easily visible and dot these same shallows here and
there, but we are so big in comparison that ever-busy aggregations of cells just look like puddles of muck to us. The smallest
object that the unaided human eye can see would be about
the size of an amoeba, just a tenth of a millimeter across. So
much of what swarms and grows and responds on Earth in
this primordial moment is many times smaller than that, only
potentially visible when all those cells group together into
something greater.

To understand this time, we need to think small. The
shift only requires a few moments and a little concentration.
As you stand in the shallows and feel the water slop and lap
against your ankles, pay attention to your breathing and imagine shrinking with every exhale, becoming tinier and tinier in

stature until you're smaller than your childhood self, smaller than a housecat, smaller than a mouse, smaller than a lightning bug, the world as you know it shifting out into an impossibly broad horizon in which even a speck of stone seems like a house-size boulder. It's so small that you could easily find the spaces between those broken-down pieces of rock and slide between them, an inch of seafloor now a vast and craggy field that feels impossibly deep despite the fact that you are still in sunlit waters. The air above the surface of the water may as well be outer space. Now you have a more fitting sense of scale, able to detect the small changes that—in time—will bring around sweeping consequences for the planet.

Shrunk down, the threads of red algae now stand taller than you are. Their full name is *Bangiomorpha pubescens*—the pubescent red algae form. It's a strange name for this organism, not least because it's difficult to imagine the strings of cells engaging in anything resembling teenage rebellion, but it's a signifier of this essential turning point in life's history. Not only can *Bangiomorpha* make its own food through photosynthesis, but it's also one of the first living things on Earth to have sex.

It certainly took life on Earth long enough to accidentally happen upon this new way of reproducing. And the shift will open up new potentials for the green life that will follow in the hundreds of millions of years to come, responsible for pollen, flowers, and life's irrepressible variety. Sex, as far as the potential for reproduction goes, is a gamble whose risks are offset by the new combinations of traits that evolution either discards or develops in the fullness of time. The path to this pivotal moment has been a winding one, in which accidents and novelty opened up new possibilities. The origin of organisms like the red algae wasn't simply a matter of slowly refining cellular processes through time, but involved global changes, mass extinction, and unintended cooperation between cells that would generate an entirely new branch of life.

Photosynthesis is not synonymous with plant life, neither in our own time or in these chilly Stenian days. In fact, many of the first photosynthesizers might not be able to survive in the world *Bangiomorpha* now inhabits, and organisms like the little red algae acquired the ability to make their own food not through step-by-step evolution but a happy accident that changed the history of Earth forever after. As we continue to watch the *Bangiomorpha* twitch to the rhythm of the sea, let's consider how such a living thing came to be.

The first cells capable of converting sunlight to energy evolved relatively soon after the origin of life. Instead of consuming other cells or organic molecules, the earliest photosynthesizers were able to use energy from sunlight to shuffle electrons from carbon molecules and other components of seawater to make sugars that nourished the cells' other processes. Those early photosynthesizers tumbling and bobbing through the water didn't give off oxygen as a byproduct, however. The cells were anaerobic photosynthesizers, meaning that oxygen played almost no role in the biological process. By accident, however, some of these ancient self-starters evolved a new way of feeding themselves that would begin to alter the Earth's composition and fundamentally reshape the planet.

As individual cells nourished themselves and split, the genetic copying and pasting carried errors with it—mutations. It's like running a passage from a book through a photocopier, scanning that copy once more, and again and again until smudges on the page appear to create a new word that changes the meaning of the page. One of these unforeseen changes altered the way some cells carried out photosynthesis, incorporating the oxygen from water molecules in the oceans—the O in H_2O—in a way that gave off O_2 molecules as a byproduct, nothing more than a buoyant gas released back into the water.

The O_2 molecules produced by a single cell would have been negligible, even over the course of the microbe's life. As the number of these oxygen-producing photosynthesizers grew, however, the chemistry of Earth began to change. Oxygen levels in the water climbed, saturating the oceans. Once the seas had become oxygenated, the gas began to escape into the atmosphere and altered the composition of the air—making oxygen molecules more prevalent than the previously abundant methane. What had once been a relatively rare molecule on the planet was produced in such vast quantities by early life that it forever altered the nature of the air and water. By a billion years before the time of *Bangiomorpha*, enough oxygen had seeped out of the ancient seas to make up about 3 percent of Earth's atmosphere. Photosynthesis had changed the planet, and not all life would thrive under these new conditions. The rise of oxygen caused a mass extinction.

On our sun-soaked planet, early photosynthesizers thrived for billions of years. The main check on the proliferation of photosynthetic cells was their exposure to the sun. Green-tinged cells clogged the shallows and turned the waters emerald where the cells thrived. The sudden surplus of oxygen, though, was toxic to many forms of the more ancient, anaerobic photosynthesizers. The photosynthesizing cells that generated oxygen molecules were killing off their neighbors, and the dead cells fell to the sea bottom day after day, year after year after year. Over time, the carbon of their bodies became incorporated into the sediment. The sediment then turned to stone, all the while the oxygen-producing photosynthesizers continued to alter the planet. Over the course of millions of years, as the rocks enriched with carbon from long-dead cells were pushed up above the surface by the movement of Earth's ever-shifting plates, something strange happened. Oxygen molecules in the air began to react with the organic carbon in the rocks, creating compounds like carbon dioxide. Oxygen levels in the air dipped, even if only temporarily. Life had caused Earth's processes to find a new balance, and the success of oxygen-producing photosynthesizers ensured that it would be a lasting one. Life did not just live on the planet. Life changed the planet. Photosynthesis changed the world.

Such sweeping alterations were spurred by the everyday activity of tiny cells. The waters of the Earth had been greened. Plants were not an inevitable consequence of the change, however. Life could have very well stayed in a single-cell state for billions of years more. What happened next was not a matter of gradual refinement that we usually associate with evolutionary change. Something much more bizarre transpired, a cohabitation that would form the basis for organisms like *Bangiomorpha* and every plant that will come to stretch toward the sun.

Among the most prolific of the oxygen-producing photosynthesizers in Earth's early days were cyanobacteria. Up close, they look like strands of green, pill-shaped cells. Their verdant tint comes from chlorophyll, which appears green to us because that's the part of the light spectrum the pigment doesn't use. Chlorophyll absorbs blue and red parts of light, leaving the green part of the spectrum to be reflected back out rather than taken into the plant's tissues. The nature of light itself set up the spread of greenery, a near-useless shade for the photosynthesizers.

The oldest cyanobacteria lived in a world of cells that could easily absorb them. Eating a cell that creates its own food is nothing but a free, extra-nutritious meal. And so cyanobacteria lived side by side with other cells that surrounded them, absorbed them, and broke them down into nourishing components that allowed other forms of life to thrive. Not every attempted meal was wholly successful. On one unknown day, a cell searching through the water column for food encountered a cyanobacteria and wrapped its microscopic body around the floating cell. Every time before, the captured cyanobacteria was broken apart into its nutritious components. Not this time. Somehow, the cyanobacteria survived inside the cell and was capable of still feeding itself. The engulfing cell couldn't bust the cell membranes to dissolve the cyanobacteria. Perhaps a prisoner, perhaps a roommate, the photosynthesizing cell was stuck inside of its host.

You can't live in such close conditions without sharing a few things. The cyanobacteria cell could still split. The host cell could divide itself, as well, and when it did so, the engulfed cyanobacteria produced a new version of itself to go along with it. A cell meant to be a meal turned into a new cellular feature, the cyanobacteria tucked inside its host cell while also providing it with food that crossed through the photosynthesizer's membranes. The cyanobacteria and host cell both had their own DNA, not packaged neatly in a nucleus but distributed like cooked noodles inside of themselves. In such close proximity, perhaps in those moments of mutual splitting, cyanobacteria genes were taken up into the core genetic storehouse of the host cell. The cyanobacteria was no longer just along for the ride. The host lineage began to reproduce the photosynthesizing cell inside of itself with every generation, two species now forever entwined as one. The new photosynthesizers were hybrid creatures, split evolutionary lineages that had come together like two streams becoming one. In time, the cyanobacteria ceased to resemble its original self and was transformed into a new feature within the parent cell. The modified version, called a chloroplast, had become a coordinated, food-generating component of its host.

The new cellular form began to thrive alongside its predecessors. The history of life isn't a story of progress, after all, but is more like a play in which new characters arrive, many depart, and some seemingly minor characters stick around through all the dramatic changes that unfold. Descendant lineages often live alongside species that very much resemble their ancestors, and so, now that we can return to the Stenian seas *Bangiomorpha* call home, we can perceive both the red algae's novelty and the persistence of the cyanobacteria.

Not far from the tiny grove of *Bangiomorpha* there are what would seem to be immense, rounded towers of compressed sediment. The structures aren't purely geological phenomena. They were created by life. Atop each one is a mat of cyanobacteria, not much different from those that now act as the food generators inside the red algae, and the forms they have inadvertently created are called stromatolites.

Stromatolites huddle shoulder to shoulder in these waters, low and flossy mounds that would seem like a distortion in the sand if there were not so many of them. Domes like these have been around since the days of the oxygen crisis, not a single organism but rather a colony that remakes the very ground beneath it. The living part of the stromatolite is a woolly green shag on top made out of cyanobacteria, an early adopter of photosynthesis. As sand particles settle among the strands or little grains become stuck in the mat, the cyanobacteria readjust to get on top of the accreting mass, pushing the particles beneath and inadvertently fixing them in place. The process is too slow to see, but as this back-and-forth takes place day in and day out for year after year, generation after generation, the cyanobacteria collect small pillows of sediment and raise themselves just that much higher toward the shine of our distant sun.

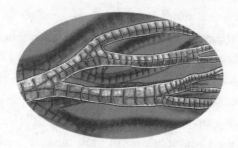

Tiny *Bangiomorpha* resembled modern red algae, protoplants with differentiated cells.

Both the stromatolites and the *Bangiomorpha* growing alongside them rely on cyanobacteria, just in different ways. While *Bangiomorpha* is not technically a plant, the threads are close to that origin point. Within the walls of each algal cell is a chloroplast, not all that different from the cyanobacteria the component evolved from. The interior of the pocket is a mess of tiny discs, little green structures called thylakoids that absorb the sunlight needed for the *Bangiomorpha* to make its own food. The plant sits near the very roots of the plant evolutionary tree, an organism that could only exist thanks to billions of years of evolution, extinction, and chance events that led some cyanobacteria to take up residence inside other cells. And perhaps *Bangiomorpha* and its relatives could have become another evolutionary blip, one form of life that came into being but did not last. Nothing is guaranteed against the unfurling of time. With our gift of hindsight, though, we can perceive *Bangiomorpha* along a fizzing evolutionary fuse that is going to set off a burst of new living things. The protoplant is not just a collection of clones, but is made up of differentiated cells in which different parts of *Bangiomorpha* carry out specific tasks.

Bangiomorpha don't just adhere themselves to the sediment like the cyanobacteria atop the stromatolites do. Superficial as their hold is, each stalk has a specialized cell that sends smaller threads into the spaces between the sand grains on the sea bottom. The fact that these holdfasts do something specific means that *Bangiomorpha* is a multicellular organism, a single living thing made up of multiple cells with different shapes and functions. The rest of the red algae exposed to the sunlight creates enough nourishment to feed the holdfast, which in turn allows *Bangiomorpha* to live in one suitably sunny place instead of being subjected to the uncertainty of floating through the water column. Perhaps it doesn't seem like it in such a minimalistic state, but this differentiation will eventually be modified and tweaked such that specialized cells grow together to make tissues that combine to create organs, and which build upon each other into systems. The world will always belong to the small single cells that can adapt to just about any habitat and can simply split to reproduce, but the origin of multicellular photosynthesizers will allow everything from ferns to orange trees to evolve.

Living in just one place comes with its own risks, of course. The holdfasts mean that *Bangiomorpha* can't move from their tiny patch of sediment. The benefit is that being anchored relatively firmly to the bottom means that *Bangiomorpha* can reach taller than many of the other photosynthesizers spreading out as mats and slimes across the sand. While something as simple as a wave carrying too much sand might mean disaster for cells growing close to the surface of the sediment, *Bangiomorpha* reach higher and have a better chance of still being exposed to the life-giving sunlight. Now that the red algae is able to grow upward, and not merely outward, *Bangiomorpha* can coexist with other species in a world crowded with constantly busy cells.

As the cloud cover begins to scatter and light shimmers here and there among the shallows, *Bangiomorpha* begin to photosynthesize once more. The threads are doing more than just making food. Differences between some of the red algae wisps indicate that there's another biological process playing out. Some *Bangiomorpha* strands seem almost twice as thick as the smallest ones. The grove is a mix of different-diameter threads, each with a different role to play in the plant's reproductive life.

For the vast majority of life's history, organisms have reproduced themselves asexually. Split with yourself and suddenly you have a slightly different copy, not a clone and not a sibling, but something more like a mild variant of the original. And it's worked out pretty well for microscopic life. Organisms can reproduce quickly and, if their constellation of traits works well in a given environment, they can rapidly spread.

Such species can also respond quickly to changes like variations in temperature or oxygen levels, as a greater number of variations increases the chance that some of the population will be capable of withstanding the latest perturbation and passing the useful traits to their offspring. It essentially comes down to numbers—if you quickly produce a vast array of slight variations at any given time, there's a better chance some will be able to survive the worst. Then again, a split can also perpetuate a vulnerability or limitation that can be difficult to evolve out of. On top of that, the amount a descendant can vary from its parent is relatively minor. Even though anaerobic photosynthesizers were numerous and reproduced rapidly, for example, many species didn't vary enough to include those that could withstand Earth's increased oxygen.

The red algae have spread by keeping tradition while adding a new twist to it. The various *Bangiomorpha* threads don't begin their lives as slightly varied copies of the original. The way they'll reproduce is more fortuitous. The stacks of cells start off as asexual, not differentiated into any reproductive role early in their lives. In time, some of the cells make asexual spores that can drift away and start the growth of a new plant much like its parent. But some of the threads change. Some cells split, and split, and split again, a thread producing two thousand or more sperm-like spores. Other spores transform in a different way, creating a kind of biological hook to gather those spores and fertilize a nucleus held within. The red algae is creating sex cells that recombine into new genetic arrangements—variations tested against the ever-shifting conditions of the planet. It's a way to introduce more variety, faster, opening new possibilities for future change. More than that, harmful mutations have less of a chance to ride along with each generation. Every time the gametes meet each other and create a new mix, there is a possibility that liabilities to survival hidden in the DNA won't be copied over. It's a different response to an ever-changing, chaotic planet. Rather than throwing out a greater number of slightly different organisms of which a relatively few might survive and keep up with ecology's pace, *Bangiomorpha* can vary more widely and have a greater likelihood that some chance change will open possibilities never seen before. In fact, it was sex that opened the possibility that *Bangiomorpha* could evolve its anchor and rise vertically from the seabed. In a rapidly changing world, sex is one way to stay ahead of tomorrow's changes.

Here, in this ancient sea in which multicellularity is still a novelty, sex has opened possibilities that will be realized through the course of time. Cells can group together as part of one living thing and take on different roles. It's no longer essential for the whole organism to copy itself. Cells can differ with each iteration, allowing the emergence of actual species with arrays of variations instead of more homogeneous groupings of near-clones that survive through sheer numbers. With each reproductive cycle, cells can potentially take on a new shape or role in the body of the entire organism, assisting in the survival of the whole. Sex requires luck, of course, and fortunes aren't always good. Some mixes of genes come to nothing. Some differentiated cells aren't particularly useful, or might even leave a living thing vulnerable. Such risks are unavoidable, yet the incredible flowering of life can only come about through such trade-offs. Sex drives such a novel spread that life could take on forms never before possible. What is simply another part of *Bangiomorpha* life, dancing to what the sea calls, will hasten how many new forms will originate and become extinct in the hundreds of millions of years to follow, a great flowering that will see life in the shallows wholly change the nature of the terrestrial realm.

2

Worts and All

425 million years ago
Oman

SAND AND WATER MAKE SUCH A WONDERFUL SOUND. Each wave draws out a gentle scour as it washes away from the tideline, as if it's the rhythmic breath of the planet itself. Not a single living thing can hear the music. The grains of the shore fade to a field of stone, rising to the naked rock of mountains in the distance. Billions of years after the origin of life, and over a hundred million years after the first animals began to float and squirm and crawl through the seas, the terrestrial realm is one largely shaped by the shifting of tectonic plates, rain showering exposed stone, sun beating upon the sand, and other abiotic influences that oversee everything from the path of pebbles downstream to the rise of mountains. Along this shore, there are no palms or coarse tufts of dune grass, and certainly not any forests in the distance. Even as life in the seas thrives and grows ever stranger, life on land keeps close to the surface. At a glance, the rocky shoreline seems almost lifeless. The incoming and outgoing waves move, but there is little else that would immediately catch our eye.

Stark as the divide between land and sea might seem, however, we can still catch hints of how life will become adjusted to terra firma. Just beyond the crest of incoming waves, scales briefly gleam above the surface in a silver flash. They are the dark and shining armor of a fleshy-finned fish thrashing to escape something larger. The fish isn't quite swift enough. A small plume of blood dissipating in the shallows represents another evolutionary thread that has ended only to feed another. The unfortunate swimmer will not leave progeny that will eventually come to know the world beyond the water, despite its relation to creatures that will one day bear claws, wings, hooves, and hands.

The snack was a sarcopterygian, a peculiar animal with aggregations of thick, chunky bones nestled within mitts of flesh rather than thin fins splayed over bony rays. Someday bones such as these will form the basis for life so varied and strange that the world will be filled with the descendants of what are, here in the Silurian, completely mundane fish.

Such changes will not come from enterprising adaptation or a curiosity of what lies beyond the crashing waves, far from the reefs and trilobites that crawl along the sandy seafloor. The fleshy-finned fish, now crammed into the dark gut of another, was well suited to the water. It had the unique advantage of breathing air when oxygen levels in the water dropped too low for its gills to efficiently breathe, and its body plan shared a great deal in common with the far-off creatures that will some-day roam dry land, but it's not as if the creature was somehow tiring of Earth's waters. The damp sand and dry soils beyond the water have nothing at all to offer such a fish. At least, not yet. Plants have to transform the land first and offer entice-ment to the animals that have proliferated through the water. In this age, greenery is only beginning to settle its roots down.

Despite initial appearances, the beach isn't barren. Every brief reach of the ocean up the banked shore washes water over tiny plants that grow in a ragged line down the sand for as far as the eye can see. The flecks of green on this beach most closely resemble what we'd recognize as liverworts, plants that live low to the ground and can only grow where moisture is assured. Each and every little bit of dark green looks as if it's stretching between two worlds, connected to the ocean but caressed by air and unfiltered sunlight. As the day shades into yet another ancient summer evening, their greenery seems slightly darker against the growing pink of the sky and the towers of puffy white clouds rising out at sea that will provide a soundtrack of thunder to the coming night. Plants, at long last, are beginning to come ashore.

Before we leave the oceans behind, however, we'd do well to remember that plants and other photosynthesizers out at sea are a large part of why Earth is a green planet as well as a blue one. Throughout time, now and constantly through the mo-ment where past shifts to the present, it's the ocean-dwelling photosynthesizers that contribute the bulk of the planet's oxy-gen, and aquatic plant life will continue to evolve just as their terrestrial relatives will. And rather than representing a single, triumphant moment, photosynthesizers have been living along the margin between water and land for over a hundred million years already, finding homes in mucky ponds and creating slimy carpets over patches of the landscape. These ancient life-forms will continue to inhabit these watery margins far into the future, too, and yet they are not the ones that will transform rocky expanses of land into forests, develop extravagant flowers, or grow into thick grasslands. While obligated to grow where it's wet, it's plants that are slowly taking root farther and farther from the waterline.

The traits of the ancient red algae have been further tweaked and tinkered with over evolutionary time to allow plants to emerge. Plants not only photosynthesize thanks to internalized cyanobacteria, but have cell walls surrounding the biological goop within and can sexually reproduce. In time, plants will expand and even defy some of these fundamentals. Some plants will capture insects or feed on the dung of animals to nourish themselves, and many that grow on other plants won't even be green, as they'll be able to draw resources from their hosts rather than making their own food. At this early stage, however, plants are beginning to grow in every suitable place their spores can reach. Through time, happenstance after happenstance have brought plants to the edge of the land.

The spread of small plants along the sand is only momentous in retrospect. We know what's to come over the following millions of years. In this time, the soaked, round-leafed little plants are simply another of life's expressions that is surviving wherever it can. The tideline plants are not colonizing. They are not invading. They are not conquering. Like all forms of life on our unusual planet, they are finding cracks in the world where they can survive—and perhaps leave something of themselves behind. Their lives are one of call and response, the twining of the living and the lifeless, water and sun creating the context for roots and green tissues that have evolved to thrive within our planet's particular atmosphere and geology. If life is a symphony, plants have begun to play an altered tune that complements the classical arrangements of the seas—a few seemingly quiet notes that animal life has already begun to join.

The shoreline liverworts, just like their photosynthesizing predecessors, produce their own food. Place them in a habitat with enough water, sunlight, and resources in the sediment, and they will thrive, able to renew their own populations season after season. The food they create for themselves, though, is just as useful to animals that must eat other organic matter to survive. In the seas, early snail-like animals evolved hardened teeth to scrape algae off rocks and sustain themselves. The existence of plants allowed for the evolution of herbivores, and, in a sense, plant predators were already waiting when plants began to grow out of the water.

About 15 million years before this Silurian summer evening, many-legged invertebrates were already crawling over the ancient dunes in search of food. Some of them looked almost like little isopods—small, segmented creatures that pattered about on spindly legs and left trails of comma-like tracks in the damp sand. Scavengers, they walked over the beaches to decomposing organic matter tossed up onto the tideline for some easy meals away from all the snatching, crushing, pulverizing claws and mouthparts that waited back in the water. Not all of these invertebrates were tiny and helpless. Sea scorpions—or eurypterids—were one of evolution's early success stories. Some of them had head shields like horse-shoe crabs, others were like their namesakes and had a more scorpion-like appearance, and many had joined arms that ended in crab-like claws. The largest of these pinching in-vertebrates could reach more than five feet in length. While not entirely at home on land, these arthropods could retain moisture on their gills for long enough to wander over the mudflats and shorelines, foraging or looking to cross overland to the next body of water. They were sluggish and could do little more than drag their bodies across the damp sand with their jointed legs, plowing shallow scrapes that were mostly smoothed back down by the next high tide. Still, there was nothing on shore that could threaten them other than the threat of drying out and leaving a cracked husk in the sun-light. None made it far inland, and yet they transcended worlds.

In the grand evolutionary scheme of things, arthropods seemed to waste little time at all before starting to explore increasingly terrestrial ways of life. Their exoskeletons and jointed legs capable of holding their bodies up off the ground allowed them to venture where other animals had not been able to follow. When all your wet, vital tissues are wrapped up in a hardened carapace that can resist drying out even for a short time, moisture caught in your gills, forays above the surface are possible in ways they are not for the fishy and squishy. A lack of fish or other predators above the surface was an added bonus, and the longer an invertebrate could stay out of the water, the better their chances of not being snaffled as another creature's breakfast. The first vertebrates to come out of the water were more than 50 million years off in the future. Especially for arthropods that survived by feeding on decomposing organic matter—be it rotting algae or the body of a trilobite tossed up onto the sand—the terrestrial world was full of nourishment. The earliest part of this evolutionary dance was regulated by chance. Decomposers fed on whatever they could find washed up based on the vicissitudes of the water and the weather. It's challenging to make a living on luck alone, nibbling at whatever the ocean has cast up onto the sand.

Now, however, greenery is growing along the yellow sand and the roving invertebrates have plenty of reason to keep returning to the world beyond the waves. Paired, parallel divots dot the sand like oversized pinpricks. They're not distributed at random, but make wandering and sweeping turns over the damp sediment. At the end of one of these tracks there glistens the carapace of a small creature not unlike a millipede, contorted into an S-like shape in profile as the arthropod leans back to rest its many-legged body on the stalk of a small plant growing along the shore. There is a vast garden to feast upon, more than they could ever eat.

The earliest land plants grew low to the ground and resembled liverworts. Their liver-like shape is what gives them their name.

The wandering arthropods are moving through groves that stay close to the ground. No greenery here grows tall enough to cast shade. The broader parts of the plant, the organism's sun catchers, can't even be called leaves. They're something more ancient and multifunctional, a sort of base for photosynthesis, a launching pad for the reproductive parts of the plant, and a broad surface for tiny, spike-like anchors called rhizoids that help keep the plant anchored to the ground—an expansive tissue technically called a thallus. Each broadened part of the plant, about the width of a pencil tip, lacks any kind of vascular system. There's no network of tiny tubes to carry water, sugars, or anything else around the plant's body. The plants lack any kind of rigid structure to reach higher into the air. It'll take tens of millions of years more for the adaptations plants gained in the water to find their full potential in the terrestrial realm. Life has to be simpler as plants grow in a realm where gravity is suddenly a much more important consideration. And that's why the little tufts of greenery are so small. Every part of the plant is effectively on its own. The parts able to get enough sun, water, and nutrients from the damp soil thrive while those that are blocked by too much of a rock's shadow, are just a touch too far from the water, or otherwise live right on that edge will wither. The error bars for survival are relatively tight so early in the story of terrestrial plants.

The small sprays of greenery along the beach add a splash of color to the neutral earth tones of the sand and the rocky expanses beyond. And as precarious as their existence might seem, plants like these will keep growing as various forms of their descendants will come and go. There is more than one way to grow outside of the water, after all. In time, the descendants of these fragile little plants will evolve ways to resist drought, attacks by insects, and other catastrophes. In some ways, plants will eventually become more dependent upon the weather and adapt themselves to the temperature and rainfall of the varied environments they encounter through time. Everything from water-conserving cactus to palm trees dripping with jungle condensation can draw their ancestry back to these little plants. For the moment, though, the liverworts don't just need water for photosynthesis and to keep from drying out under the harsh sun. The plants need to live in the damp in order to reproduce.

Jutting up from one liverwort, the one suffering the depredations of the voracious little millipede, are projections that look like tiny trees. Each is just a pinprick compared to the thallus that it's growing from, opening up into a tiny umbrella-like structure. On this particular plant, the structure is an antheridium—little storehouses of sperm that are about to play the constant evolutionary game of chance. Each of the sperm have two whip-like extensions sticking out the back, flagella that the released sperm will twitch and flick as they move over the thinnest veneer of water in the sightless push to find their reproductive counterparts. Because on some of the neighboring liverworts, jutting up in very similar structures, are archegonia. Tucked inside the upside-down-teardrop shape made by those very small cell walls are the egg cells.

After emerging from the waves, the liverworts split their lives into parts. Whether they make sperm or eggs, each plant has a half of the species' completement. It's only when those gametes meet that the plant develops embryos that kick off an entirely different part of the life cycle, the chromosomes split again into two parts to send out into the environment as spores. The plants don't have control over where those spores land. If too many egg-producing plants land near each other, they might not get fertilized. If the spores land in dried-out patches of the beach, the liverworts won't grow there. The fate of the spores depends almost entirely on wind and water, sand and sun, which then grow just to add more chance into the process. It's a wonder that the plants are able to connect at all, their lives broken out into separate pieces reliant on one risky circumstance after another. And yet they're here, giving the beach a hazy green shade all along its length. In the mornings and evenings, with the sun low in the sky highlighting the bottoms of the clouds, the waxy exteriors of those innumerable thalluses seem to glow in the orange-and-golden light. The greening of the Earth is beginning in earnest.

New opportunities can also generate unexpected consequences. Every living thing that has ever existed, lives in this moment, or will ever come into being has a direct relationship with Earth's atmosphere, climate, hydrological cycle, and all the other nonliving aspects of the planet. In turn, life changes the conditions of the planet to alter everything from the global climate to the erosion of stone. The divide between the biotic and abiotic doesn't mean there is no relationship between the two realms. It's not a matter of life *on* Earth so much as life enmeshed with Earth, and plants have taken an ever-greater role in shaping the planet's history. And just as photosynthesizing cyanobacteria once triggered a mass extinction of organisms that eschewed oxygen and struggled in the chilly times an oxygenated atmosphere helped perpetuate, the plants that are rooting themselves on land are once again nudging conditions on Earth in a different direction.

The plucky little liverworts, doused by the waves that leave them glittering in the setting sun's light, are creating a new ice age.

Even though cyanobacteria, red algae, and other photosynthesizers have been pumping out oxygen for over half a billion years by this point, Earth's oxygen levels have remained relatively low compared to those that will prevail during human history more than 440 million years after our beachside scene. Earth's atmosphere began with little more than a trace of oxygen, so the oxygen crises of earlier eras weren't created by excesses of oxygen but a rise to about 3 percent of the atmosphere's composition. That's far below the twenty-first-century levels of about 21 percent. And it's this moment, as plants are coming ashore, that atmospheric oxygen is about to undergo a huge jump. The change will affect much more than the lives of the little plants and the animals that feed upon them.

Above the surface of the seas, the amount of oxygen in this prehistoric atmosphere is about 16 percent. It's not simply a matter of plants exhaling increased O_2. Plants and other oceanic photosynthesizers still contribute much more oxygen to the air than the relatively few plants and cyanobacteria that have been able to live in terrestrial and freshwater habitats. If the influx of oxygen was just coming from the shore-dwelling liverworts, it would be much less. Instead, it's the death of the plants that has been giving the world's oxygen levels a bigger boost.

Plants build their bodies with carbon. It's built directly into photosynthesis, taking carbon dioxide and using the sun's energy to split up the carbon and oxygen—utilizing the first internally and releasing the other. The reaction alters what happens to those elements in the atmosphere. When storm surges cause waves to scour vast amounts of sand from the tide break, spewing sediment all over liverworts on the shore, the plants can't photosynthesize and perish in the damp dark. In their immediate burial, however, oxygen in the atmosphere doesn't recombine with the carbon in the plant tissues. Largely shielded from atmospheric oxygen, the plants don't decompose because oxygen in the air can't recombine with the carbon of their bodies, and so the carbon is held within the Earth. The carbon-rich sand is eventually compressed into stone as sediments stack over millions of years, while oxygen levels get a relative boost. What the liverworts are doing, without any awareness of it, is taking carbon out of their environment only to be buried so quickly that the oxygen that might otherwise lead to their decay remains in the atmosphere. Repeated over and over and over again, oxygen levels rise because plants grow in habitats where they're likely to be subsumed and made part of the Earth's rock record.

Increased oxygen in the air can create an entire cascade of changes. For one thing, there's now enough oxygen in the atmosphere for fires to sustain themselves. Fire can't do much more than smoke and smolder when atmospheric oxygen levels are below 10 percent, as they were for much of Earth's history. And while there are not yet forests to catch aflame, plants are slowly generating that possibility for themselves as oxygen levels creep higher. The way some plants will eventually evolve to live with fire, and even require fire as part of their life cycle, is a change of their own making.

The plants are even changing the sediments and rocks they grow among. Terrestrial plants don't simply send their anchors down into open space and shove in among all the small particles as if they were plastic aquarium decorations. Plants can weather down elements like calcium and phosphorous among the substrates they live on. The plants are breaking away components of the geological world they're entwined with, allowing those elements to concentrate in new ways and further altering the nature of life on Earth. Calcium, for example, is an important component of bone, itself a relatively new biological tissue that some animals have become reliant on just as plants are beginning to settle in onshore. Plants take calcium out of the rocks and into themselves, and when rains wash dead plants or sediment created from root-pulverized rocks back into the sea, that calcium becomes available as raw materials for fish to turn flexible skeletons made of cartilage into more rigid bone. The fleshy-finned fish that was munched along the shore was one of these, its bony skeleton at least partially owing itself to the calcium land plants are taking out of the rock so that it can be redistributed into the water. The liverworts are transforming the planet with no idea of the large-scale changes that are already underway, shifts to the way elements like carbon, oxygen, and calcium cycle through the planet's abiotic systems. And in a world where living things have a new hunger for carbon dioxide, even the climate will change.

The greenery constantly doused by the salty waves took in its carbon from the air in the form of carbon dioxide. It's the stuff plants build their bodies from in the same way our species will someday use wood to make our homes. The molecule vastly affects the climate, with higher levels contributing to warmer global temperatures. Now, however, the molecule is being sucked up both by plants waving in the waters and the greenery popping up on land. The increased number of plants soaking up carbon dioxide is causing the molecule's presence in the atmosphere to drop, diminishing its warming effects. As plants continue to grow on land, global temperatures will dip by about nine degrees Fahrenheit on average, enough to cause miniature ice ages that affect life both in the seas and trying to get a hold on land. Climate isn't something that happens to life, but is part of a never-ending feedback loop between Earth's biotic and abiotic parts.

Earth is transforming faster now. The planet's fate isn't merely shaped by the composition of its elements, the rock cycle, or the chunks of space rock that occasionally fall to the planet. Life is molding the planet and its fate as much as plate tectonics and ocean currents. And as life does so, it's opening pathways for what might come next. Evolution, after all, is an unavoidable consequence of life's present diversity meeting the new challenges of the next moment. Chance mutations and adaptations honed over millions of years are constantly being tested and affected by new and ever-changing conditions. Now that green plants can survive farther and farther away from the water's edge, all the traits they are bringing with them will set the basis for something that has never existed on the planet before. Soon, the first forests will reach tall into the air and provide both food and shelter to the creatures that flop, slither, and crawl out of the water.

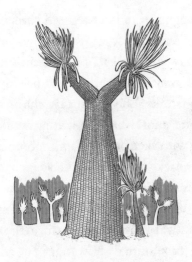

3

The Forest Primeval

307 million years ago
Ohio

THERE ARE NO SONGS IN THE NIGHT. NO CHIRPS, NO chitters, no arthropod string section ready to give the repeating darkness its rhythm. Now and then something splashes in the turbid water, or there's the scratch of a jointed leg on bark, but the dark hours are not yet filled with the various croaks, trills, and screeches that indicate how thoroughly *alive* forests are. It will happen in time, tens of millions of years from now. But in this moment, this time, life on land is still novel. Forests themselves are still something new. The land has only just begun to roll out a green carpet tall enough to shade, envelop, and enclose.

The soft glow of a full moon causes innumerable little spikes and rough pieces of bark to cast prickly shadows over countless tree trunks. The botanical columns grow tall and thick here, some more than six feet around, vast sprays of fronds jostling with each other for unimpeded sunlight during the day. Plants pack into every available inch, a dense, warm, and humid forest where the millipedes grow more than six feet long and alligator-size amphibians silently watch the shoreline for unwary insects. And the arthropods flourish. They were the first animals to wander out of oceanic waves to seek out the slimy shore plants that huddled close to the sand, hard exoskeletons made of chitin allowing them to carry their moisture inside while our fishy vertebrate ancestors were still tens of millions of years away from setting fingers and toes on shore. And here, among these woods, arthropods grow big.

Reflected light glows along the edges of a dark carapace, dozens of intricate exoskeletal pieces making up the back of what at first seems to be a moving part of the forest floor. Never before, and never again, will an arthropod as large as *Arthropleura* crawl over the Earth. The invertebrate is the largest land-dwelling arthropod of all time, a creature that can only exist in a forest of such extravagant greenery. The entire animal stretches more than eight feet from the front of its head to the trailing tip of its many-segmented body. Despite its intimidating size, however, it's truly only the plants of this Carboniferous forest that have anything at all to fear from the animal. *Arthropleura* is an herbivore, its complex mouthparts best suited to munching through increasingly tough plant parts. The evolution of such a truly giant plant predator was spurred by the origin of forests capable of reaching far higher into the air than ever would have been possible for the shoreline liverworts of a hundred million years earlier. The key is lignin.

Plants didn't evolve lignin in response to coming ashore. The biological compound had evolved in other circumstances and been co-opted to new uses. Evolution excels at refashioning what already exists into new biological possibilities. Hundreds of millions of years before this moment, in the single-celled world that makes up so much of life's history, various photosynthesizing organisms evolved the precursors of lignin in the form of monomers—molecules that are like links that can be chained together into longer chains called polymers. Lignin monomers were already present in the red algae that wafted among the cyanobacteria, not to mention other cellular organisms that rely on photosynthesis like diatoms and dinoflagellates. It didn't allow algae to grow twenty feet tall, but instead helped protect such cells from the UV radiation of the sunlight they so desperately needed to expose themselves to in order to photosynthesize while they spread through the sunlit shadows. Plants evolved their own sunscreen. And if early plants and their predecessors had continued to explore the life aquatic, the practically weightless environment may not have given lignin much of a nudge to become such a vital support structure. As plants began to establish themselves on solid ground, however, lignin began to change how plants grew.

Linked together as a polymer, lignin is very strong. That's

a critical consideration in how high plants can grow above the ground. Without something to strengthen its cell walls, a tree without lignin would bend and flop over under its own weight like a stalk of week-old celery. Plants would have stayed low-growing, not even high enough to cast much shade on the ground. Earth would not have forest but endless shrubbery, at best. Lignin provided plant cells with the necessary strength to begin growing tall. But that's hardly all. Lignin is also important in water transport, allowing plants to soak up water through their roots and move the liquid through the tree's tissues, reaching higher into the air than would otherwise be possible. By pure evolutionary accident, a molecule that had evolved to help protect plants from the damaging effects of UV also included hidden potentials that trees needed to grow far above the surface of the soil, providing both strength and a way to move water from the ground into even the tallest parts of the tree.

This forest primeval has its own strange character. The trees of this Carboniferous forest aren't anything like the forms that will thrive over the next 300 million years. Each towering trunk is immediately recognizable as a tree, but not quite like any that will flourish during the heyday of the dinosaurs or shade the proliferation of Cenozoic mammals. The shape of these plants identifies them as trees, but in future times their surviving relatives will return to growing low to the ground rather than reaching for the distant stars overhead.

Forests are composed of trees, and a tree is not a distinct group of living things but truly just one expression of what plants can become. In fact, the difference between a tree and a shrub often rests on how tall the mature plant can grow. Extend a shrub farther into the air and it gains a new title, just as a tree that branches low and has little trunk to speak of could easily be counted as a shrub. Shifting between such botanical expressions will happen again and again among different plant groups through time. Even though they're more closely related to ferns and other ancient plant forms, the scale trees of this forest are trees because they are tree-shaped, just as the pines, palms, and peach trees that will evolve later are trees despite belonging to distantly related plant groups. In this particular scene, when the possibilities of what a "forest" might be are centered around the biology and adaptive potentials of the plants that were tightly connected to the water, trees are effectively giant liverworts and ferns, immense and sturdy versions of plants that will make up parts of the forest understory as time continues to unfold second by second. Within these swamps, where plants run riot during a time when the average annual temperature is a humid sixty-eight degrees Fahrenheit, there is one tree that is the icon of the time—*Sigillaria*.

When liverworts and other early land plants started to take hold among the sand and rocks of ancient beaches, plants couldn't even shade the invertebrates that came ashore to feed upon them. But now moonlight slips through thin, blade-like leaves held over ninety feet in the air. Waving in the breeze, the leaves of *Sigillaria* create a blanket beneath which the rest of Carboniferous life carries out its nocturnal activities. Not bad for what is essentially a giant moss.

In the dark, groves of *Sigillaria* look like clumps of thick, tall trunks opening into sprays of drooping leaves near the top. The lunar light that shines between them highlights the rough texture of their trunks, almost scale-like. The tree made the texture as it's grown, scars of years past visible on the outside. If you were to cut into one of the gently swaying trunks, you wouldn't find wood or neat rings inside. The core of each *Sigillaria* tree is pith, a spongy tissue that will later form an important component of plant stems. Around that pith are a kind of plant connective tissue, called leaf bases, where the actual leaves attach to the trunk. It's almost as if the entire tree is one oversized leaf on an oversized twig, and all the scaly bumps and divots along the outside are the old attachment sites for the leaves that now grow like a shock wig at the top of the tree. The arrangement seems awfully minimal, especially for a plant so large, but *Sigillaria* doesn't rely on its leaves alone to photosynthesize. When the sun rises in a few hours, the trunks will seem green, photosynthetic tissue spread around the tree's circumference to give it some extra surface area to make its food.

Sigillaria isn't the only giant creating Carboniferous arbors. The coal swamps are places where terrestrial life of various forms can finally grow to sizes that were previously impossible, not so much creating a kingdom of giants as expanding the range of how big it's possible to be. Giant insects, giant amphibians, giant spiders, and other giant plants all thrive around the base of the tree, as well as other oversized species. Its relative *Lepidodendron* can sometimes grow even taller than *Sigillaria* does, not to mention the enormous horsetail *Calamites*, tiers of rough foliage growing from near the ground to over one hundred feet in the air. Plants that had previously huddled low to the water now grow as tall as a seven-story building. Lignin allowed these plants to rise so tall, and a warm, wet climate has created such favorable conditions that more plants are growing than decaying.

Plants are becoming increasingly important to the foundations of Earth's ecosystems, but it's not as if they lack relationships with other forms of life. Bacteria and fungi—the world's great microbial clean-up crews—have evolved and adapted right alongside these incredible trees. When one of these trees perishes or is pushed over by winds too strong for the scaly bark and pith to withstand, they crash to the mucky ground with a great susurration of their wispy leaves that sends small, lizard-like creatures scurrying. From there, bacteria and fungi go to work almost immediately to eat away at the plant tissues—from the entangled, spreading root systems to the vibrantly green leaves. As soon as there's a vacancy, new plant life begins to grow in the same plot, reaching toward the sunlight as small animals shelter within the enclosed space of vacated trunks. The forest is made of both the living and dead to change the habitat faster than perhaps any other time in Earth's history. *Lepidodendron*, *Sigillaria*, and their neighbors are growing faster than dead plants are decomposing. If ever there was an Age of Plants, this would be it.

So many plants are growing that they are leaving a permanent mark on Earth's rocks. The fossil record has been growing for as long as there has been life, a product of happenstance in places where sediment is being laid down rather than scraped away. The Carboniferous forests often grow in such places—swamps where streams, seasonal floods, and even heavy rains can draw mud and sand over so much of the carbon locked up in the dead and decaying plants. Buried, surrounded by sediment turned to stone, and gradually shoved into the Earth's crust by the slow push of plate tectonics, the rotting forests will eventually become massive coal seams that will one day be released back into the atmosphere to alter future climates as greenhouse gases, the legacy of Carboniferous trees burned up more than 307 million years after they met the sunlight. The glut of greenery is shaping other lives, too. Plants have always been food, and new animals are beginning to take advantage of such generous salad bars. Even lizard-like mammal predecessors are switching from a life of nabbing insects to awkwardly mash fronds and leaves in their peg-toothed jaws.

In these days, among the sweltering swamps, our ancestors looked like reptiles. Called synapsids, the crawling, scaly creatures are new arrivals on the ecological scene. To tell them apart from the ancestors of dinosaurs, crocodiles, and lizards among their neighbors, you'd have to look inside their heads to the back of their skulls. The early reptiles thriving in these same wetlands usually have two openings at the back of their skulls—one on the side and one above—that mark their storied lineages as diapsids. Over the millions of years to come, these early reptiles will spin off lizards, dinosaurs, pterosaurs, crocodiles, and similar scaly creatures throughout time. But synapsids only have a single opening in their skulls along the side wall, what will eventually become the opening just inside the arch of the primate cheekbone hundreds of millions of years from this waterlogged moment. For now, the synapsids—more simply, protomammals—are beginning to grow into their major evolutionary moment that will see them thrive for the next 50 million years.

Despite their lizard-like appearance, *Melanedaphodon* are more closely related to us than reptiles.

Sails are in fashion among many of these beast precursors. Low on the forest floor, shuffling along with a splayed posture and scaly belly near the shadowy ground, is a little *Melanedaphodon*. The protomammal is no titan. The entire animal is less than two feet long, much of that being tail. Still, it'd be hard to miss this hungry little *Melanedaphodon*. Jutting from the green-tinted synapsid's back is a low fan of thin skin hoisted over elongated spines from the creature's back vertebrae, each one with tiny, spike-like crossbars running along their length. Later relatives of this protomammal will grow much larger, becoming some of the first big herbivores the Earth will ever see, but *Melanedaphodon* is only just beginning the tradition. When plants grow thick, some animals give up on hunting to chew the scenery.

The teeth jutting from the small protomammal's mouth

are not suited to slicing or breaking bone. The rounded, enamel-covered pegs are the sign of a plant eater who doesn't mind some crunchy insects along with their forage, an omnivore teetering on the edge of herbivore. Not that he's particularly good at it. He can't chew. His teeth evolved to puncture, crack, and slice, but not grind. Nor does he have the jaw musculature to pulverize and mash vegetation across the teeth like duck-billed dinosaurs or cows will do many millions of years from now. The overall effect, as he tucks into a low-growing horsetail, is as if he's smacking his scaly lips, chlorophyll and pulped vegetation giving his reptilian smirk a greenish tinge as small shreds of plant land damply onto the mud below.

To compensate for an inability to chew, protomammals like *Melanedaphodon* evolved another way to make feeding on fibrous plants easier. The roof of his mouth is dotted with tiny teeth, an extra set of mashers to break up insect shells and fibrous stems before going down the hatch. And he needs to keep smacking his jaws through most of the day. The advantage of being an herbivore is that there's food almost everywhere. The disadvantage is that tough greenery is much more difficult for the body to process, often requiring substantial modifications to make the most of every meal. Bodies convert flesh to vital tissues much more easily than plants. Over the course of evolutionary time, herbivores will repeatedly evolve ways to more efficiently grind their food, like broader guts to house digestive systems that worked like fermenting vats, and even hosting a symbiotic relationship with bacteria capable of releasing energy from otherwise indigestible plant food. It's an ecological balancing act that will cause carnivores to become herbivores and herbivores to become more carnivorous time and again.

For the moment, *Melanedaphodon* splits the difference. There's no shortage of green food for the sail-backed protomammal, but the squat, shuffling creature won't turn up its nose at insects, either. And in these forests, the insects grow big. *Arthropleura*, sliding among the downed tree trunks, is by far the largest, but it is not the only supersized arthropod here. During the daylight hours, dragonflies with wingspans two feet across flit and buzz through the mazes of green trunks.

The dragonflies didn't get so big out of some evolutionary drive toward ever-larger sizes. It's the explosion of plant life that's underwritten their evolution, especially the products of photosynthesis. The spread of plant life all around the planet, forming multi-tiered forests full of greenery, has increased the amount of oxygen in the air to an incredible degree. While Earth started out with almost no oxygen in its atmosphere at all, the success of the plants has created a new atmospheric balance with about 35 percent oxygen—the highest level of all time.

On the surface, extra oxygen might seem like an advantage to organisms that evolved to require the molecule for their survival. The salamander-like amphibians living among these coal swamps breathe through their skin. Higher oxygen content in the air means that they can more easily take in oxygen and their body processes can work more efficiently, opening up the possibility of some impressive, squishy-skinned creatures in the ancient mire. Some of Earth's swamps host creatures like *Crassigyrinus*, an eel-like amphibian that could grow to be more than seven feet long—the descendant of amphibians that went back to the water and became more at home in the swamps than the protomammal and reptile lineages that are beginning to emerge on land. And arthropods have benefitted from the oxygen spike, too. Dragonflies and other arthropods breathe through trachea, a network of tubes connected to shutter-like openings called spiracles. More oxygen in the air means it's easier for adult insects to breathe more efficiently and nourish their tissues at larger sizes, allowing their evolution to test the limits of how big exoskeletal organisms can get before they become too big to support themselves. Bigger arthropods would have greater interior volumes compared to their outer surface area, their exoskeletons imposing a limit on how big millipedes, dragonflies, and other many-legged creatures can get before literally bursting. Without this constraint, surrounded by food and assisted by higher oxygen levels, arthropods may have grown to become even larger still.

Then again, excess oxygen creates new problems even as it opens new possibilities. The dose makes the poison, and too much oxygen can lead to a form of poisoning, even affecting the central nervous system of animals. Larger size is one way to compensate, a further nudge to the impressive sizes of many coal swamp creatures. By starting big and growing larger, in other words, animals are able to skirt around some of oxygen's drawbacks. Among the swamps and puddles of the Carboniferous forests, larval insects have had to evolve to contend with the new world that plants have created.

In the dark shallows of a pond ringed with ferns, dragonfly larvae crawl among the algae-covered bark of scale trees that have toppled over each other. It's a vegetative snarl that is a little too tangled for many larger organisms to poke their noses into, and thus they shelter in a relatively small space. The young insects are somewhat sensitive to the higher oxygen levels in the water, fostered by the higher concentrations in the air, their small bodies vulnerable to the negative effects of increased oxygen.

The larval insects can do practically nothing to change the high oxygen levels the plants of their world are creating. Nor is there anywhere to move to where conditions might be better. The only routes open to them have been to adapt or go extinct. Luckily for the arthropods, the puzzle can be solved by becoming bigger. Larger animals aren't as sensitive to oxygen overload as smaller ones, and so, generation by generation, slightly larger dragonfly larvae have fared better than their smaller kin that were more prone to oxygen's toxic side effects. Bigger larvae opened the possibility of larger adults during a time when higher oxygen content could actually be an advantage for active flying insects buzzing through these ancient forests. Incredibly elevated oxygen was not a gift given to life on Earth, but a condition that created an evolutionary puzzle that favored some unprecedented, enormous animals.

If the flowering of terrestrial life on Earth can be expressed musically, the interaction between animals and plants has been relatively simple until now. Now the music is beginning to swell into more of an orchestra. On land, forests have multiple layers to them: groves where herbivorous creatures are beginning to gain new ways of biting into the fibrous fronds and breaking down those tough cell walls. Oxygen levels will climb and fall as climates shift, each a back-and-forth between how plants themselves respond to the world around them and how they shape those non-living forces.

What seemed so straightforward, even elegant, is beginning to erupt in a cacophony of evolution and extinction, a ceaseless tune in which new players will imperceptibly, fundamentally change the melody with each passing moment.

4

Fire and Water

220 million years ago
New Mexico

THE BEETLE IS AN ABSOLUTE JEWEL OF A MORSEL, ORANGE light of the early morning glinting off the arthropod's jet-black shell. The warm hues filtering through the trees are about as much light as this part of the forest is going to get, golden rays spreading out low to the horizon before everything below the riot of branches throw purple-and-indigo shadows over the ground cradling the forest. Moving slowly, dotted with tiny spheres of dew, the beetle shifts one leg after another as it walks along the fronds of a low-growing fern in search of a smaller insect to munch on. And just like that, both beetle and half the fern are gone, rapidly scissored apart in a reptilian mouth brimming with leaf-shaped teeth. A multi-jointed beetle leg tumbles to the forest floor, left to time or some other creature small enough to appreciate the leftovers.

Silesaurus doesn't need to stalk his prey. Crunchy bugs may fly off if the movement of his shadow gives him away first, but, most of the time, the German Shepherd–size reptile is so big as to just be another part of the scenery to the little insects. And in these ancient woods, there are plenty of crunchy snacks. One bite after another, the lanky little omnivore fills his warm belly with broken carapaces and segmented bodies mixed with Triassic greens. It takes many such bites to fuel an endothermic metabolism, after all, but at least all the fatty, crunchy beetles seem to give the wispy, dark feathers covering his neck and back a soft sheen of their own.

Fifteen million years earlier, *Silesaurus* would have been an evolutionary novelty. He's close to dinosaurs, but not quite a dinosaur—a creature left over from the great evolutionary boom that followed Earth's last mass extinction. It was a particularly bad one. The Triassic was borne of a volcano-driven disaster that undercut life for more than a million years. It was a slow, grinding disaster, ecological pressure so great that more species were going extinct than were coming into existence. Excesses of carbon dioxide and methane pumped out of suppurating volcanoes, turning oceans acidic, pushing the global climate to unbearably hot, changing ocean currents, and even reducing the amount of oxygen in the air. More than 75 percent of all species on land and in the seas perished, and those that were able to survive had to make a living in a world where entire food webs continued to collapse even as they began to reestablish themselves. It's almost as if Earth temporarily took on the attributes of an alien planet, one overly hostile to the life it had helped shape.

Entire families of living things were decimated. Protomammals were incredibly varied and widespread before the disaster, evolving into iguana-like leaf-eaters, weasel-like herbivores, and saber-toothed carnivores as large as a black bear. Of all their number, only two lineages managed to survive—squat little mammal predecessors called cynodonts and tubby, tusked dicynodonts that munched on low-growing plants with toothless beaks. Somehow, however, reptiles were able to hang on where the protomammals slipped. Many evolved in new ways to cope with the chaotic world, assisted by reaching sexual maturity early and laying clutches of multiple eggs. The surviving reptiles reproduced at such a rate that they evolved to fill the Triassic world faster than the surviving protomammals, not only evolving into new forms on land but also sliding into the seas in a rebuke of harsh terrestrial habitats. The protodinosaurs were among them, *Silesaurus* a walking monument to the success of a particularly diverse group called archosaurs.

As he struts through the morning forest, the occasional squawk of a toothy pterosaur breaking the quiet, he carries on the legacy of the protodinosaurs. The ancestors of the first "terrible lizards" looked very much like him—not slobbering, ravenous, knife-toothed carnivores ready to sink their claws into the world, but gangly omnivores that munched on fern fronds and insects. The earliest dinosaurs didn't rule anything. The first of their family led a relatively retiring existence on the margins of Pangaea's ecosystems for millions of years, filling ecological cracks between the greater array of crocodile cousins that filled the world faster. Dinosaur forerunners—what will be dubbed dinosauriformes more than 220 million years from this time—persisted alongside their bizarre offshoots, an earlier form somehow surviving alongside creatures that are novel among the forests, floodplains, and deserts of the Triassic.

Little by little, the hungry *Silesaurus* daintily picks his way through a forest of towering conifers and ferns speckled with tiny orbs of dew. Horsetails grow thick around a nearby pond, though going for a drink in these woodlands can be deadly. Even though all the crocodiles run around on land—some even walking on two legs, as the dinosaurs do—evolution has a persistent fondness for flat, trap-jawed creatures that can conceal themselves in only a few inches of water. It's the inappropriately named phytosaurs that are the biggest dangers here, some of them exceeding twenty feet in length. More than once the *Silesaurus* dipped his jaw into the shallows to drink, feet still planted on the sodden soil, when a great toothy rush would come at him in a spray of water. You can't drink here without being ready to run, one eye on the surface of the water. Even the smallest and seemingly safe puddle could conceal a monster capable of puncturing down to the bone.

The forest provides plenty of places for such knife-toothed predators to hide. Trees don't rise straight into the air, opening into shaggy leaves at the top like in the coal swamps of the Carboniferous. Here, the core of the forest is *Araucarioxylon*, a cousin of cycads, ginkgoes, and conifers that sends out a spray of branches from bottom to top. The forest is a mess of long, nearly naked branches ending in tufts of almost needlelike leaves. The tallest among these trees grow more than two hundred feet straight up from the forest floor and measure almost ten feet at their base, held firmly into the soil by a dense cluster of roots that can reach more than fifteen feet down into the soft earth. The trees don't need thick bark to withstand strong shifts between the seasons. It's hot here— the lower temperatures a cozy 68 degrees and the warmest days reaching over 120 degrees of dry, unrelenting heat that make dinosaurs pant, open-mouthed, to keep cool. Today is one of the hot ones, already over 80 degrees as the sun is just beginning its climb up the sky.

At least all those wet, sloppy parts inside the insects and arachnids offer some hydration for the *Silesaurus*. The silesaur isn't quite parched yet, ignoring the small trickle of a streambed as he wets his clawed feet in the flow. Thunder booms low in the distance, seeming almost comforting as the sound unfolds its bass notes across the landscape. A few little skips and he's up the bank, dainty scaled feet pressing a little deeper into the ground. Behind him, towering forest has been huddling close for century after century, long before he kicked his way out of his egg within its confines. But before him is a razed and scorched landscape of cracked trunks, toppled trees, and charcoal.

The dry seasons here are merciless. Even the menacing phytosaurs, seemingly impervious to any injury from their neighbors, sometimes become stranded in ever-shrinking, noxious ponds where fishy prey is driven ever closer to them until there is one last, mucky meal as the last of the water is taken into the air. Those that don't leave become macabre monuments of cracked bones and bleached jaws spattered by rains that came too late. The weather won't return what it's taken for months to come, and in the interim fire takes a turn at shaping the landscape. Desiccated needles and branches litter the forest floor, piling up as if in anticipation of the conflagration that both destroys and renews. Plants did not evolve to prevent forest fire. They evolved along with the consequence of their own being, living according to a primordial cycle spread out over a timeline most animals don't live long enough to even perceive.

Many of the fossil logs in Petrified Forest National Park and other Triassic fossil sites are what remain of pine-like *Araucarioxylon*.

Above the ground, *Araucarioxylon* isn't especially resilient to fire. The trees grow close in a hot, relatively dry climate, the soil soaking up maybe twenty inches of rain altogether in a good year. When lightning sparks dry forest litter, huge swaths of forest can burn. Each year, some unlucky patch of Triassic forest is left as little more than a smoldering patch of blackened wood riddled with small fissures. But a tree's life is not just its conspicuous existence above the surface. For *Araucarioxylon* to grow so tall, they require strong, spreading roots. Those roots wind their way deep into the soil, protected from the flames of the scouring fires above. No matter how vicious a conflagration it is, perhaps rising over five hundred degrees Fahrenheit as the oxygen created by the plants feeds the flame blasts that consume their tissues, the heat of those fires rapidly dissipates within the soil. Less than a foot beneath the surface, it may as well be that there's no fire at all. And it's these immense taproots that allow *Araucarioxylon* to keep pushing new sprouts out of ashen fields, reestablishing the woodlands where insects bore into their wood and dinosaurs yawn languidly in the shade. Life can hang on by a matter of inches.

Hunting's good on the flats expanding from the opposite bank of the silesaur's preferred foraging grounds. None of the plants among the scorched patches has grown back higher than the reptile's head. All the bugs hiding among the vegetation are scuttling close to the ground, and he nabs whatever he can find. He weaves between the burnt stumps and over fallen trees that have not yet been ground back into the earth, his long tail gently bobbing as he wanders. There are bones here, too. The fire had burned so fast, and so hungrily, that many could not escape the smoke and the walls of flame. Blistered and burnt bodies reduced to a few bones, here and there contorted and cracked skeletons of what were once lithe dinosaurs, deep-jawed crocodiles that lunged from behind the trees, and chameleon-like reptiles that clambered hand-over-hand in a world held entirely aloft in the canopy. They are too dry to even gnaw, their skin, muscle, and marrow lost long ago. Only the bones speak to the lives that once filled these woods.

A few fat drops of rain thump onto the ground, and one even catches the reptile flat on his head between his eyes, making him tilt his chin up at the darkening skies. Intense yellow light flashes one last time as the sun climbs high enough to get above the cloud cover, leaving an ominous and dark blanket of a building storm. Without a sound or any other perceptible cue, the arthropod buffet seems to disappear. All the *Silesaurus* can find is the butt-end of beetles tucking into the bark and the occasional spider running up the slope, just barely escaping the ravenous reptile. Too soft for him to hear, a great curtain of rain begins to approach the sandy incline he's been standing on. He can still smell it, though, a change in the air as the scent of rain-driven dust sweeps over the parched ground. *Silesaurus* chatters at the intense interruption to his errand, turning tail back toward the shelter of the gently rocking conifers across the stream. The storm's faster, though, the pelting drops soon overtaking him and coming down with an intensity that would almost sting if the sensations against his scales weren't so overwhelming.

The protodinosaur picks up the pace, no longer trotting on tiptoe but running full-out through the dendrological graveyard. The ground can't handle the downpour. The previous fire destabilized the entire stream bank, dried-out soils so thirsty for moisture that they soak the rainfall up too fast. The silesaur's feet start to slip a little on the saturated ground, fine particles of ash behaving like slippery clay beneath his clawed feet. What he meant to be a firm and fleeting step turns into a skid, all of his momentum suddenly turning against him as he slips in the gray muck and topples to his side, flank knocking against a coarse and blackened stump before he crashes to the ground with a surprised screech. He flails and rights himself, half running and half stumbling back down toward the stream that's beginning to look fuller and darker than when he had crossed. The running water is a soupy gray, a tea made of water, ash, and burnt debris. His splayed back feet pound the streambed and splash the thickened water back up, dappling his messy flanks even more, as he reaches the opposite bank just as it gives way.

There's too much water. Rain can only be incorporated into the earth so fast, and the dried-out, ashen ground lacks the community of plants, fungi, and other living things that hold the soil together as water drips down. The ground is quickly oversaturated, heavy with water that is taking too long to absorb. Gravity soon takes over. A great chunk of the bank—gloppy three-toed tracks of the *Silesaurus* still visible on it—suddenly collapses and rushes down toward the stream. Pieces of wood, charcoal, bones, and innumerable tiny lives are all carried with the flow, broken apart and sloshed in the fall as what the fire left behind is mixed together. The break causes other patches to slip, as well, the entire hillside reshaped in mere moments as the silesaur finds an accommodating tree hollow to curl up in and watch the rain drum the forest floor with a tense eye.

Some of the terrible slush spills along the stream channel, a messy flow that's both solid and liquid that rushes along at nine feet per second. Even creatures that were far from the storm will be swept up in the mess, the force of the slide taking miles to expend itself. But not everything is dispersed over the miles. The initial slump started to create a chaotic pileup, the beginning of a literal logjam that will collect pebbles, boulders, logs, and bones as it settles. It's as if the landscape had just been run through a huge, disjointed sieve, collecting tidbits of what the fire didn't consume.

In time, the immediacy of the catastrophe abates. The muck still acts like a trap for days after. Even as the upper muddy layers begin to dry and remain solid, the sucking mire beneath remains liquid. A shuffling *Desmatosuchus*—a snouty crocodile relative that looks like a reptilian armadillo with great spikes jutting out above the shoulders—wanders too far and sinks armpit-deep into the trap. The hapless creature's honks of distress lure a carnivore, the tyrannosaur-like *Postosuchus*, that only manages a single bite into the herbivore's armored back before becoming stuck alongside. Dehydration under the hot sun is inescapable. Neither of the reptiles enters the fossil record. They're too exposed; bugs and bacteria break down their bodies and the new route of the stream eventually washes their battered remains downstream. But below, safeguarded by the sudden rush of mud, is a woody time capsule of this Triassic world.

You would think that a tree, safely snug and buried below the surface, would have a simple and straightforward afterlife. The once-thirsty wood would seem all too eager to slurp up the mineral-laden water and transform. But like many things trees do, the afterlives of these gymnosperms play out over unfamiliar timescales.

The bark of the Triassic tree did its job. Patches and parts were burnt and weakened by the fires, but the internal tissues of the tree remained intact. The burnt outer rind decays away to expose the still-resilient, relatively untouched wood

within. Tissues that were once bound up in the living tree
are exposed within the sodden ground, refreshed now and
then by rain that plinks its way through the soil and stream
water that slips below the rivulet's belly to the hidden treasure trove. Little by little, what was once wood is replaced. In
time, the trunk will become petrified wood.

Every living thing on Earth is composed of elements and
compounds from the Earth itself. The hard, mineral part of
the bones of the silesaur waiting out the storm in the relative
safety of the forest is made of calcium carbonate, a compound
that erodes from rocks to become dissolved in water and taken
up into bodies that have evolved to incorporate the material
into stiff internal scaffolding that wouldn't have been possible
without the mineral's widespread availability. The tree, in its
own way, had built itself from the carbon it took in from the
air during its daily photosynthesis. And now this tree will be
welcomed into the fossiliferous afterlife because of another
mineral, silica, carried through the trickles and surges of the
stream that buried it.

About 10 percent of the Earth's crust is silica, from
quartz crystals ground down to specks of sand and to meta-
morphosed portions of the Earth's crust. Water dissolves
and carries silica easily and abundantly, even more in places
where there's a greater concentration of volcanic glass or ash
slowly being eroded back down—the insides of the Earth
being reshuffled into parts of the outer layer.

The transformation isn't something that can be seen directly, and certainly not when the Triassic tree is already buried beneath the sand of the streambed. But under the blanket
of sediment, in the dark, silica-laden water trickles through
the tiny spaces between sand grains and bathes the tree trunk

with the literal ebb and flow of the stream. Season by season and year by year, the trunk is bathed in fresh water that carries an invisible mineral complement within it, bathing the log's sponge-like interior with the suspended minerals. The hydrogen of the suspended silica grabs on to the oxygen of the cellulose making up the tree's cell walls, creating a thin film of silica that acts as a kind of primer for even more silica to adhere to the quartz rind building up in the tree's tissues. As the process is repeated through the years, the silica starts to build up within the microscopic cell walls as well as the outside, the comparatively heavy accumulations settling to the bottom and packing so tight to become opalized in a rainbow of colors. The nature of the stone forming within the framework of the tree depends on the surrounding rock and history of submersion, a record of how it was formed as unique as the tree's life. There's only a limited amount of time that a tree can undergo such a transformation. As resilient as bark and wood are, they are not immortal. Silica has to pass over the wood often enough, in enough of a concentration, to be welcomed in by the tree's tissues before decay recycles the tree back down into its molecular components.

The water trickling through the tree's tissues carries more than just silica. Silica is the key component in preserving the tree, reworking the tissues from something organic into something mineral, but other geologic drifters make themselves at home, too. Iron oxides and quartz are soaked up into the wood, all of which will give the preserved wood striking hues of red, yellow, blue, and smoky purple, a rainbow hidden within the rock. Over the coming eons, if the rock is in just the right place, nudged in just the right way, and gently scraped by water and wind, those colors might get a chance to shine in the daylight—a rainbow that rises from the stone.

An abridged story of the tree's life is enclosed within its transformed tissues. The flow of wet seasons to dry and back again are recorded in the width of the rings emanating out from the log's core. The wood's rainbow colors owe themselves to the minerals saturating the groundwater where it once grew and was buried. Trails made in the wood by burrowing insects are preserved, too, ephemeral moments of time that are fossils on fossils. The tree was built from the elements around it and transformed by the very earth it was once rooted to, no longer the original but something beautifully transformed, able to last lifetimes longer than the plant ever experienced in its own. Painted in glorious shades, the fossil tree has become a memory held safe within the rock until it may one day be discovered and asked what life was like so very long ago.

5

Land of Giants

150 million years ago
Utah

SHE'S BEEN GRAZING FOR HOURS BUT IT'S NOT ENOUGH. There's still an emptiness in her great reptilian belly. After all, sixty feet and twenty tons of dinosaur is quite a lot of bone and muscle to feed, all of her busy tissues requiring energy from the vast spread of plants that clothes the stream bank in vibrant green as she munches away. The sun is shining low in the sky, the air has cooled off from its midday peak, and there's nary an *Allosaurus* to spoil her breakfast. Perfect. Head low to the ground, she opens her modest muzzle once more—

the cropped frond of a horsetail falling from the side of her mouth—to shear off another mouthful of Jurassic greens with her single row of pencil-shaped teeth.

In a far distant time, paleontologists will name her species *Apatosaurus excelsus*. She is a sauropod, an especially thick and burly member of a group of long-necked, plant-eating dinosaurs that will include the largest creatures to ever inhabit our planet. The fact that she exists at all is a testament to the strange nature of her habitat, a vast woodland where stands of towering conifers and monkey puzzle trees huddle together among seas of ferns and cycads whose fronds waft in the wind like waves. It's a landscape of the very big and very small, of giants too big to be fussed by the lack of cover and skittering creatures tiny enough to hide among the low ground cover. *Apatosaurus*, naturally, falls among the giant category, and to her most of this landscape is potential food. Most delicious of all are the carpets of *Equisetum* that grow along the streams and oxbow bends on ancient Utah's Late Jurassic floodplain, horsetails growing from the muddy margins 100 million years before horses will even exist.

From a distance, the groves of three-foot-high horsetails look like three-foot-high shag carpet. Up close, they're more like the bottle-brushed tail of a cat, just bright green and banded from tip to base. These are very ancient plants. The first *Equisetum* evolved about 50 million years before our present moment of Jurassic time, around the same time that dinosaurs were staggering out of the world's fourth mass extinction. They've changed relatively little since that time. Each horsetail is a hollow tube, rising from a network of roots below the surface that connect these thick stands of greenery. That central stalk does all the photosynthesis, held upright by the support of many, many cell walls that hold the energy-producing organelles safely inside. The brushes at the top, despite their color, contribute little to the plant's near-constant food-producing process. These wisps are not so much leaves as proto-leaves, simple filaments that are to a leaf what the single filament shape of a dinosaur's protofeather is to later broad plumage. As simple as they may seem, though, these plants have done very well for themselves—quite a feat in a place that dries out for months a year.

The Late Jurassic world is warm and humid. The average annual temperature hovers around sixty-six degrees Fahrenheit, warm enough that a temporary Sundance Sea has covered much of prehistoric Canada and coral reefs accrete from the equator to zones that in later times will be considered temperate. Million years after million years, the world feels in the midst of an endless summer. But the consequences of global climate are not uniform around the planet. The Earth is warm, and relatively wet, but where that life-giving moisture's distributed depends on currents in the sea and air, the rise and fall of mountains, and the innumerable facets of natural history that leave some places dry while others are absolutely drenched. For the organisms that call the lands of the Morrison Formation home, existence is tied to the arrival and departure of rain.

These Morrison Formation lands are inland, far from the coasts of ancient North America. The moisture that rises from the Pacific Ocean is blown east by the prevailing winds, feeding rain clouds that darken the skies and drop torrents as they travel toward the interior. Their path is obstructed. Little by little, tectonic forces are pushing up mountains to the west of the Morrison Formation basins, snagging some of the moisture—water that slowly erodes the rock even as the mountains rise—that would otherwise make it to the stomping grounds of *Stegosaurus* and neighbors. By the time these nourishing storms reach the lands of *Apatosaurus*, they have largely been rained out. It's only during the monsoon season that the rains are strong enough to return and refresh the Morrison Formation lands, leaving its inverse—a harsh dry season—the other half of the year. The dinosaurs that live here do what they can to cope with these seasonal swings. Some stay and seek out wetlands and sources of fresh water as streams dry up and some ponds turn into alkaline mires that are nauseating to drink from. Other dinosaurs move out toward the coast, basins and floodplains kept just the much wetter by the proximity to those evaporation-fed rainclouds. Of course, the problem is much different for *Equisetum*. The plants stay put, their roots anchoring them in place. New stands of horsetails pop up every wet season, growing as sodden opportunity allows, but only those fortunate enough to be anchored in moist soils along the margins of rivers and lakes remain year in and year out, the moist, crisp staple that *Apatosaurus* and other herbivores rely upon. Through the energy they convert from the sun, these plants are a power-packed lunch that have helped underwrite the expansive evolution of some of the largest animals to have ever trod upon the planet.

Our hungry, hungry *Apatosaurus* was a recipient of this ecological largesse before she even hatched. Her mother, just like her mother before her, munched on the horsetails, fiddleheads, cycad fronds, conifer branches, and other fibrous foods common to her patch of western North America. Those plants allowed her to grow and to survive, one day after the next, from her terrifying days as a tiny pipsqueak hiding among the ferns and other low-growing foliage—so small she'd barely be a mouthful to a hungry *Ceratosaurus*—through adulthood, her daily diet permanently staining her saurian lips a rich green. She reached sexual maturity long before she came close to attaining her full stature, the plants that fermented in her gut unlocking the energy that would allow her fertilized eggs to form inside her body. This wasn't just a matter of genetic switches being thrown off and on or dictating her life's trajectory. Genetics alone are useless. DNA only makes sense in the context of its environment, an interplay between what genes can tell a body to do and whether that body is having its requirements met. Her body, down to her chromosomes, held so much potential, possibilities that could only be unlocked through her daily visits to the Jurassic salad bar.

Now her daughter—the only one of a dozen hatchlings to survive the violent and harsh first year most dinosaurs face post-hatching—calmly stands in a clearing along the edge of an oxbow bend in a meandering stream, sunlight warming the scales and wispy filaments of her striped back. Those lighter marks, thin lightning bolts of cream against the burnt orange of her flanks, neck, and tail, are all that remain of her camouflage from when she was a youngster. During her first year, those stripes looked larger and broke up her outline as she foraged and hid among low-lying plants that seemed to form their own kind of forest over her tiny head. Ferns like the huge, broad *Angiopteris* were once her shelter, helping to conceal her from the battalions of carnivores that share these forests and floodplains: *Torvosaurus*, *Allosaurus*, *Ceratosaurus*, *Marshosaurus*, *Stokesosaurus*, *Coelurus*, and more, all with their own needs and hungers and roles to play in the Mesozoic drama. With curved talons and serrated teeth never far away, it's a wonder she survived at all.

The only way to survive was to get big—and fast. Baby *Apatosaurus* hatch out of grapefruit-sized eggs and are but a tiny fraction of their potential size. Within a year, however, these dinosaurs can grow larger than a horse. Those lucky enough to survive a few years more match or exceed the maximum lengths of their predators, often hanging out in small herds to better spot danger—and reduce the chances of being singled out by particularly stealthy assassins. Even then, an adolescent *Apatosaurus* is so burly that a well-placed tail swing could bruise, batter, and break the bones of an attacker, and a prone *Allosaurus*—our *Apatosaurus* had learned during one particularly messy experience—could be easily stomped into a scaly pancake. After that, sauropod dinosaurs like her were too large for any but the most desperate predators to harass. For all their ferocity, the likes of *Ceratosaurus* had delicate bones full of hollows and always walked on two legs. Something as simple as a fractured shin could be a death sentence if it prevented a meat-eater from effectively hunting, and in carnivorous calculations, big, healthy *Apatosaurus* are among the most dangerous of prey.

She's still wary. While she's too large to be topped by the menacing meat-eaters, our *Apatosaurus* surely doesn't need a bite taken out of her hindquarters when she's not looking. A grove of monkey puzzle trees, their branches spreading wide like an upside-down umbrella, grow on the opposite bank of the stream, but behind her is moist, trodden ground, too open to let any carnivore sneak up on her unawares. With her eyes situated on either side of her head, she can keep watch out while she grazes, taking in about ten inches of horsetail with every bite. Each mouthful has a pleasant snap, not unlike celery, as she feeds, swallowing and never chewing.

During her younger and more tender years, she shared the vegetative spread with others of her species. More eyes mean a better chance to spot an approaching predator and less of a chance of you being pounced on. But that's hardly all. There are benefits beyond protection for young dinosaurs that gather together. Being an herbivore is a tough gig. The bodies of big herbivores, like *Apatosaurus*, can't make the necessary enzymes to break down the tough cell walls of all the tasty vegetation in this place. The dinosaurs rely on microscopic roommates to internally prepare their meals. The digestive system of our *Apatosaurus* is lavishly populated by bacteria capable of breaking down plant cell walls, releasing the energy stored inside in the process. It's an evolutionary dance that neither partner intended to start—as if they could even know—but that's become essential to the survival of both over time.

If *Apatosaurus* cared for their offspring after hatching, then the dinosaur before us may have received an essential array of plant-busting bacteria directly from her mother. An affectionate nuzzle from a mouth can pass on more than warm feelings. Sauropods, however, are not doting parents. While some dinosaurs lay relatively few eggs and watch over those nestlings, at least long enough so that those infants can feed themselves, the giant long-necks have wound up with a different parenting style—which is none at all. Gravid mothers gather together each spring, creating vast nesting grounds where they dig shallow divots in the earth to nestle and warm their eggs. Once those eggs are buried, their work is done. Each mother wanders off to feed and fend for herself, perhaps unaware of whether she'll ever recognize her own offspring should they someday meet. Scavengers and predators know this. The nests are shallow enough to be accessible to maws and paws, with carnivorous dinosaurs, mammals, snakes, and crocodiles coming back again and again to pluck an egg or two from the ground. Survival is a numbers game, the persistence of *Apatosaurus* reliant upon stuffing predators with so many eggs that some are allowed to gestate to term undisturbed. Those babies that hatch then wander out into the world, wide-eyed into the relative shelter of the forests. And this is where they receive their bacterial gift.

No digestive system is 100 percent efficient, especially among warm-blooded animals like the dinosaurs. Not to mention that plant food is difficult to break down, especially the tough and fibrous parts like branches. When adult *Apatosaurus* leave a warm, green pat on the Jurassic soil, that jettisoned material is full of undigested plant material and bacteria. Food is food, even if it's been run through a dinosaur first, and so baby *Apatosaurus* often eat from the plops left behind by the grown-ups, receiving the essential bacteria they'll need to keep the process going. They even share the gift. Hanging out together in little crèches, dodging the sharp-toothed shadows, the baby dinosaurs touch bodies, nip each other, and otherwise jostle each other enough that every one of them eventually ingests the essential bacteria that then spread through their digestive systems. The process is heavy on luck and happenstance, but, for millions of years, it's worked.

Our *Apatosaurus* is large enough now that she need not be concerned with finding a herd. Sometimes other dinosaurs will graze nearby. Little *Dryosaurus*, two-legged herbivores that nip plants with their beaks and crunch insects with their cheek teeth, sometimes gambol and forage around her column-like legs, picking up whatever plants she half chews and nabbing the insects disturbed by her steps. A plate-backed, spike-tailed *Stegosaurus* will sometimes join her as well, keeping an herbivore's accord to maintain a comfortable distance while small pterosaurs land upon their great backs and pick up blood-sucking insects from between their scales. But she doesn't seek out company. If anything, grazing with others of her stature—or larger—might mean more competition for the vegetation she relies on to survive.

On the surface, foraging isn't a particularly difficult task for *Apatosaurus* or most of the other sauropod species that share these floodplains. The dinosaurs can feed high and low, from the ground into the trees, and their long, muscular necks give them quite a reach. All she has to do is sniff out a suitable patch of greenery, stand somewhere in the middle of it, and slowly munch her way through, tilting her neck just slightly left or right, back or forward, as she works through the available undergrowth. She doesn't chew even a mouthful. That's because she doesn't have molars, nor the complex jaw apparatus that some later dinosaurs will evolve. Her jaws open and close, her peg-shaped teeth restricted to the front of her sensitive muzzle. Snip and swallow, snip and swallow, all day long. She is essentially a sixty-foot-long, twenty-ton vacuum, horfing down mouthful after mouthful. All of that intact plant matter passes down the great length of her esophagus to her stomach where acids and muscular movements of her digestive system just barely start to break down the phytological meals—as well as whatever insects, worms, dead mammals, or other morsels of protein she takes in by accident. This is all merely preparation for a long, protracted journey through her hindgut, specialized portions of intestine that slowly break down the messy plant material and extract as much nutrition as possible. Her guts are effectively enormous fermentation vats, which also results in no short supply of vegetal sauropod farts. All together, in fact, the methane output of herbivores like her are great enough to make the global climate slightly warmer than it would otherwise be, helping to sustain the seemingly endless summer of the dinosaurs.

Our *Apatosaurus* can only swallow so many mouthfuls of horsetails before the forage starts to feel a little mundane, however. She lifts her head high, above the level of her shoulders, and turns to the scene behind her. In the distance, a lone *Supersaurus*—over one hundred feet of sauropod, the largest creature in this basin—slowly ambles along, presumably to their own preferred feeding spot. In the air, a pair of leathery-winged pterosaurs squabble at a larger, solitary pterosaur of another species, dipping and diving and mobbing to move the intruder away from a nest. Furthest of all, just visible by the contrast against the dark trunks of a conifer stand, a lone *Allosaurus* throws back their head and yawns, jaw opening nearly ninety degrees before snapping shut. The carnivore rolls to the side, raising a three-clawed foot to scratch at the scales and fluff along their neck—a predator at rest, perhaps still working on digesting last night's meal, and too far away to worry about. Slowly, muscle group after muscle group tensing and relaxing to lift her legs, the *Apatosaurus* turns and begins to amble toward a grove of ginkgoes, her head and neck slightly bobbing as she goes, the tip of her whiplash tail wafting and gently flicking as if she were an enormous reptilian feline.

Ginkgoes are even older than the horsetails. The first of their great family evolved about 120 million years before our current tableau, in the days when protomammals were the most diverse and noticeable animals on land. These trees have survived through two of the world's worst mass extinctions, but not just by sheer luck.

Earth is a chaotic planet. Catastrophe doesn't always come in the form of a shuddering volcanic outpouring or asteroid crash-landing in the Earth's rocky skin. There are disasters and disturbances that are much more common, even mundane, like damp stream banks collapsing into the water or monsoon floods sweeping over the flats. In the environs of the ancient Morrison Formation, these ecological accidents and happenstances are simply part of what makes this place so unique. There are months that seem parched and dry, when ponds and streams are reduced to puddles—or may disappear altogether. Then, by the time the bones of dead dinosaurs begin to crack underneath the sun, distant thunder begins to sound and torrents of water dropped by gravid clouds race over the thirsty ground, often piling the bones and carcasses of the deceased—from the smallest freshwater clam to the biggest sauropods—into massive bone beds, gathered and entombed all at once. And in such a place, defined by the ebb and flow of water, ginkgoes thrive.

There are several species among the forests here, rising up from moist soil near the more permanent aquatic features of the landscape. The trees *Sphenobaiera* and *Czekanowskia* have long and thin leaves that almost look like feathers, their delicate lobes fluttering and wafting in the breeze like the plumage of a small dinosaur. But our *Apatosaurus* is slowly wandering toward a different ginkgo, *Ginkgoites*, flush with finger-like leaves that form a crescent-shaped spray of foliage at the end of each twig. There's an entire grove of them, thick trees with twisting branches growing from thick trunks into a maze of wrinkled bark. What a perfect meal for a hungry *Apatosaurus*.

To a carnivore, the green expanses of the Morrison ecosystem are little more than setting. A tree can provide shade or cover to stalk prey, a dry branch can give away the step of a hunter in the dark, and most low-lying plants are simply underfoot. If a horn-faced *Ceratosaurus* could process the thought, the predator might assume that one plant might be as good as any other for forage and that herbivores have it easy. The reality for herbivores is quite different and requires something of a discerning palate. Each plant—and each plant species—varies in its benefits and costs to plant eaters, especially as they get big.

There are monkey puzzle trees all around the Morrison ecosystem, their bristly arms splayed high in the air. Most herbivores can't reach the foliage they offer, and even those that can—like the impressively tall sauropod *Brachiosaurus*—may find the spiky vegetation hard to digest. Dinosaurs keep elevated body temperatures and quick-running metabolisms, which is not always an advantage for an herbivore who benefits from keeping vegetation moving slowly through the digestive system so that it can ferment and release as much energy as possible for the dinosaur's big body to use. The needles of monkey puzzle trees are so resilient to breakdown that it takes about three days inside a dinosaur's digestive system before most of the nutrition in those needles is released and can be used by the body. That's a long time for a dinosaur's digestive system to hold on to food, which is even more difficult for warm-blooded animals or even those that may be dealing with the unfortunate consequences of dino diarrhea. But there are alternatives, and ginkgoes have become sauropod staples.

Step by step, our *Apatosaurus* comes close enough to the nearest ginkgo in the stand, stopping near enough that she could touch her muzzle to the tree trunk if she wished to. But scraping bark from trees—while a last resort in meager times—is very time-consuming and not very worthwhile. Bark is like biological cardboard: rough and lacking in what her body's craving. It's the cell walls of the leaves, after all, that hold so much of what she needs. She opens her muzzle and takes a big bite, nearly a foot of leaf and twig shoved into her maw, and clips off the first mouthful with teeth beveled to little edges by all the plant food she's been stuffing herself with. She swallows and the plant material, tiny insects still clinging to some of the leaves, begins its journey down twenty-five feet of dinosaurian neck, still traveling as she takes another bite, pulling her head back to snap the twig off and making the entire branch shake as she pulls her meal free.

There is so much to eat here. *Ginkgoites* leaves grow at a greater density on each branch than other gymnosperms in the Morrison. For the biological effort of each bite and swallow, *Apatosaurus* is able to get more food from a ginkgo than a monkey puzzle tree or even a stand of horsetails. More than that, these leaves are rich in protein. That's hard to obtain for an herbivore. *Apatosaurus* gets a little protein from the insects and fungus she inadvertently consumes with her greens. In fact, she once was eating so hastily that she accidentally sucked down a small mammal that had been hiding in the branches—a most unpleasant swallow she didn't wish to repeat. But these supplements are hardly enough, and they were especially paltry back during her teenage growth spurt. During her teenage years she put on about 3,500 pounds each year, literal tons of flesh and bone that were fed by her herbivorous diet. Protein-rich plants like ginkgoes helped her to do so, and sauropods who grow up in habitats lacking in ginkgoes sometimes wind up smaller unless they can otherwise find the needed biomolecules from another source. A ginkgo might not seem all that different from the various other conifer-like trees that have sprung up around these floodplains, but they're critical food for these dinosaurs.

Multituberculates thrived during the Jurassic, squirrel-like mammals that could gnaw on seeds as well as dinosaur carcasses.

All this shaking has upset one of the tree's inhabitants. A ball of angry gray-and-white fur charges down the ginkgo trunk and toward the muzzle of the *Apatosaurus*—*Ctenacodon*, a multituberculate. These mammals are the squirrels of their time, their name derived from the almost ludicrously complex array of cusps on their cheek teeth. There's little these mammals don't do. Multis gnaw on dinosaur bones, they snack on insects, and, in the case of the tiny beast reading the riot act to the sauropod for disturbing his slumber, they are more than capable of gnawing into seeds and nuts.

Ginkgo fruits stink. Actually, they positively reek, and mature trees are often replete with these little stink bombs. And that's an enticement to the animals that will help the tree reseed this season and the next. The rotting stench of ginkgo fruits attracts plant predators much smaller than *Apatosaurus*; pterosaurs sometimes pluck the fruits from the branches, and some multituberculates can't resist the carrion aroma. Sometimes the scent of death isn't a warning but the sign of a free meal. Of course, this is a gamble for the tree. Multis often eat through the fleshy outer covering and break the seeds inside, destroying the hope each capsule represents for the ginkgo. But the mammals and other fruit-hungry organisms ingest the fruits just often enough to unwittingly help the trees spread, the swallowed seeds eventually deposited in new places with a little gift of excrement. The trees themselves can't move from where they are planted, but their assistants have helped *Ginkgoites* and its relatives spread far and wide.

Apatosaurus is unfazed by the charge, partially because the dinosaur can't see straight ahead and the sounds of the angrily chattering mammal are too high-pitched for the saurian to hear. The dinosaur opens her mouth for another helping of ginkgo leaves, the approaching maw sending the *Ctena-codon* scratching and scrambling back into the branches and around to the other side of the trunk where the irate fuzzball can shout his protest in relative safety, his tail poofed out like a horsetail frond. Best not to get in the way of a hungry dinosaur. In fact, our *Apatosaurus* is eating so many ginkgo leaves that a hole in the tree's cloak of green is beginning to open. A single dinosaur can strip most of a tree's vegetation if they really settle in, leaving the tree with nothing more than damage to recover from. But just as ginkgoes have evolved to thrive in disturbed habitats, they've also evolved to withstand the dinosaurian onslaught.

At the base of the *Ginkgoites*, just visible above the level of the Jurassic soil, the tree's bark seems swollen. The base of the trunk flares out to a collar of bark—a lignotuber. Inside, protected behind the rough bark and too low down to even be noticed by most creatures, is a biological investment. The lignotuber is full of high-energy starch and dormant buds, still connected to the root system that the tree has already grown. If our *Apatosaurus* were to be joined by some of her neighbors and their foraging were great enough to kill the tree, if a forest fire ripped through this woodland, or if tussling dinosaurs even felled the tree with all their thrashing around, the buds within the lignotuber would be able to grow anew—faster, in fact, than the tree originally grew in the first place. In a place as dinosaur-churned as the Morrison, where the stomping of huge feet and the swaying of muscular tails can be as much danger to plants as ravenous bellies, lignotubers ensure that groves remain steadfast even after catastrophe. The trees themselves have no knowledge of this strategy. They have simply evolved to respond to environmental chaos over the tens of millions of years they've already existed. It's these hidden abilities that will see them through while other forms of life come and go. Even when the time of the great sauropods ends, there will still be ginkgoes for millions of years more. If a tree can survive the Age of Dinosaurs, there is little such a resilient plant can't endure.

6

In Bloom

125 million years ago
Northeastern China

THE SMALL BRUSH OF A PLANT SHIFTS WITH THE RIPPLES, each ring in the water pushing the stem just a little more toward the nearby shoreline before it settles back into place. The aquatic swaying doesn't hurt the pipe cleaner–like stems. They simply wave back and forth with the motion as easily as a fern or a leaf might lilt in a light breeze. The shovel-beaked dinosaur creating the disturbance with each step of its pillar-like feet, however, is another matter altogether. Measuring fifteen feet from the tip of her tail to the end of her rough, beak-covered muzzle, this *Jinzhousaurus* is busily munching on the greenery growing along the lake margin, her broad back feet leaving mushy, three-toed imprints in the silt that encloses the roots of the aquatic plants.

There's never been an herbivore quite like her before. Even though vertebrates of one sort or another have been dining on plants for over 200 million years, at least since *Melanedaphodon* was munching through the coal swamp undergrowth, most of them have greedily sucked down as many leaves and shoots and stems as they could without anything even resembling a chew. Herbivorous dinosaurs, too, have traditionally stuffed themselves with unchewed greenery that is pulped and broken down within their guts, their mouths acting as little more than anatomical salad tongs. This hungry *Jinzhousaurus*, though, levers her lower jaw up and down, the bones of her skull flexing back and forth to mash the sloppy mouthfuls of vegetation she's busily grazing upon. With each chaw, leaves and stems are mashed across her huge teeth, ground to chlorophyll-stained pulp that turns her mouth green.

She isn't aware of her unusual skill any more than she's aware of her own evolutionary history, and the brushy water plant *Jinzhousaurus* is happily feeding upon is also something relatively novel. Her afternoon snack is *Archaefructus*, one of the world's first angiosperms.

In the context of this time and place, among the streams and forests of ancient China's Yixian Formation ecosystems, *Archaefructus* isn't any more important than any other plant growing on the planet. There's no hint of destiny about the growths or what they are eventually going to become. It's only hindsight that allows us to watch the small and delicate plants wafting in the shallows and know that they mark the humble beginnings of an important lineage, plants that will eventually reshape the world in new ways—from tropical jungles to dry grasslands spreading between ragged mountains. But not yet. The Early Cretaceous world is still one where conifers, cycads, ferns, and horsetails make up a great deal of the world's greenery, and the great evolutionary profusion of angiosperms is still a long way off. The first angiosperms might appear as little more than scrawny weeds in a botanical world centered around storied and well-established plants.

To us, the word "angiosperm" is more or less synonymous with "flowering plant." In later times, angiosperms are going to be the plants that burst with incredibly colorful, smelly, pollen-dense flowers that are critical to their life cycles. But, as with all things biological, the truth is far more complex. Angiosperms are not the first plants to grow flowers. Not by a long shot. As far back as the Permian Period, before the origin of the first dinosaurs, brushy plants called bennettitales grew their own versions of flowers, and these plants were more closely related to conifers than *Archaefructus* growing in Early Cretaceous ponds. Angiosperms will eventually produce a greater array of flowers than any other plant group, but not just yet. The plants are growing in whatever cracks in the world they can find between the gymnosperms, ferns, and other forms of plant life that have already been established for tens of millions of years.

A return to the water is one way to get around the pervasive hold of the conifers and cycads. In fact, it's often when living things cross into a new environmental boundary— returning to the water, flying into the air, or venturing into other wide-open spaces—that they rapidly diversify and become more significant parts of their home habitats. While there are plenty of conifers in mucky places like floodplains and swamps, relatively few could truly be called aquatic. Even then, the few pines that grow in the water do best in running water—not the calmer settings of ponds and lakes. Gymnosperms became so thoroughly suited to the terrestrial realm that *Archaefructus* was able to find a comfortable niche in the water, even if it means that many of them are easy snacks for plant-eating dinosaurs that venture down to the water for a drink.

Over the course of time, plants made it possible for there to be plant predators. Hot-blooded metabolisms require a great deal of energy to run, and plants have provided a significant portion of what makes the scaly and feathery saurian menagerie possible. Ever since plants began coming ashore more than 280 million years earlier, they've restructured Earth's food webs. Plants can make their own food, taking care of themselves with enough water, light, and carbon dioxide in the air. Animals have taken advantage of their abilities, evolving ever more intricate ways of breaking down plant matter into energy that fuels their bodies—as well as those of the carnivores that prey upon them. What plants provide is the foundation for so much life, the base on which animal life rests. And in this evolving relationship, animals have unintentionally offered their assistance to plants, as well.

Not far from where *Jinzhousaurus* is peacefully grazing among the water plants, concealed among the branches of a tree with roughly spade-shaped leaves, is a feathery little dinosaur that paleontologists will later name *Jeholornis*. It's about the size of a crow but still has a great deal in common with the feathery, raptor-like dinosaurs the first birds split off from about 25 million years before this moment in time. *Jeholornis* still has a long, bony tail ending in a splash of feathers, and the bird is picking at the leaves of the tree with tiny, blunted teeth. Despite its predatory ancestry, the bird's lineage has settled into a more herbivorous lifestyle over time, picking at the fruits of trees and munching on the green leaves despite lacking the impressive jaws of *Jinzhousaurus* below. And this particular tree that our *Jeholornis* is nipping at is another angiosperm, a very ancient predecessor of magnolias.

The primordial magnolia is just what *Jeholornis* has been looking for. Squatting on a branch, letting its tail droop below and digging its sharp little toe claws into the bark, the bird nips at cone-like fruits hidden among the leaves. Left to themselves, tiny seeds will eventually pop out of the cones and drop to the ground to take their chances on the forest floor. It's a gamble for the parent tree, one that's repeated the same time each year and relies on a great deal of chance. This forest not only grows thick, with plenty of other plant species trying to root in whatever open ground that might open up, but the woods of the Yixian are brimming with all sorts of life big and small. There are the more conspicuous non-avian dinosaurs, of course, but also pterosaurs, birds, and a variety of small mammals. The small beasts, in particular, have evolved amazing dental toolkits of sharp incisors and crushing molars, teeth capable of busting open the seeds and erasing their chance of passing through. Trees-to-be compete with each other for space, and even if they begin to grow they may quickly be snapped up before they get a chance to grow tall. It's a chaotic, undirected form of pruning that alters entire landscapes. Being eaten isn't all bad, though. For the ancient magnolia, the arrival of *Jeholornis* is very fortunate. The fact that the dinosaur can't chew will give the seeds it's swallowing a chance.

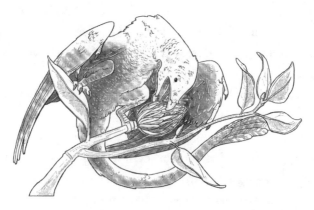

By eating and evacuating seeds, early birds like *Jeholornis* helped Cretaceous forests grow.

Back in the Jurassic, the ancestors of *Jeholornis* were consummate carnivores. They pounced on small prey and pinned it down with curved killing claws held on the second toe of each foot, scratching after mammals in their burrows and snatching crunchy insects out of the air. When birds split off to become airborne, they left hypercarnivory behind. They needed their arms to flap and clamber around ancient trees, not catch prey. And given that flight favors small size, tackling sizable prey became less and less of an option. Early birds shifted their diets to small lizards, insects, carrion, and, naturally, plants. Of course, they still had to operate within the constraints that their ancestors had given to them. *Jeholornis* and other early birds are best suited to plucking food and swallowing it, nipping morsels and jerking their heads to

throw food to the back of their throat. Whatever they swallow is processed inside, leaving the more resilient pieces to
eventually pass through virtually unscathed. And considering the competition for insects, small reptiles, and mammals
scurrying through the trees from other early birds and pterosaurs, *Jeholornis* is taking the long-standing offer of herbivory.

Living off plants requires a whole new set of adaptations.
Leaves, fruits, and seeds don't break down easily, especially
when you're relatively small. The hot-blooded metabolism required to power rapidly beating wings, too, means that the
plant food is going to spend a relatively short time in the
guts of the early bird, and so every bit of caloric energy that
can be drawn from the plant material matters. The bird can't
be a hindgut fermenter like *Apatosaurus*, trying to squeeze
resources out of plants over days and days. So *Jeholornis* swallows pebbles. They're not altogether hard to find, especially
as the streams and waterways of the Yixian roll and transport
tiny stones as they flow along. All *Jeholornis* has to do is hop
down to a stream bank now and then to pluck up just the right
water-smoothed stones and swallow them whole, letting the
rocks sit in his belly for a while. When he swallows fruits and
leaves, the rocks act as a set of stomach teeth that mash the
plant material up and create more surface area for the bird's
digestive acids to break down the plant material and provide
the energy the fluttering little dinosaur needs.

The process isn't entirely efficient, and that's part of what
makes the bird's relationship with its favorite tree work. As
Jeholornis picks and pecks at the magnolia fruit, many of
the seeds moving through his throat and to his gut will be
busted and broken down by the stomach stones. Some won't.
Some seeds will manage to ride above the grinding mass

and survive their residence inside the dinosaur without being ground down into bits, moving through his intestines and eventually excreted with a fresh dollop of dung. And this is what will allow the magnolias, and other early angiosperms, to start spreading. When the tree merely drops their seeds to the ground, their progeny are trying to grow in the parent tree's shadow. Any seed will have a vanishingly small chance of finding just the right spot to get enough water and sunlight to grow, and that's not even considering how many would-be trees are gobbled up by *Jinzhousaurus* and other herbivores when they're nothing more than tender little shoots. Seeds carried inside *Jeholornis* can be transported much farther away, to shorelines and patches of forest that might have a little more potential. The odds are still against any particular seed finding just the right spot to germinate and grow to maturity, but the survival of the magnolias isn't reliant on all or even half their seeds finding those perfect spots. Only a few need to make it to start spreading seeds of their own, gaining ground through their unique relationship with the dinosaurs. A small assist from *Jeholornis* gives the magnolia a slight advantage over trees that rely on the kindness of the elements to nurture the next generation.

Neither the bird nor the tree intended this relationship. The bird is merely looking for food to eat, an herbivore living within a realm of seemingly endless meal options. The tree spends its entire life in one place. The fact that a bird can take the magnolia's seeds and bring them to new habitats is an improvised relationship between organisms that have not shared a common ancestor in billions of years. Something as simple and direct as hunger can help establish entire forests, a close connection in which fulfilling a basic need and a little luck can help groves of trees spring up from the earth. The trees of these constantly changing forests produce oxygen, can be eaten as food, offer microhabitats where all manner of creatures live, and act as starting points for innumerable small interactions between distantly related species. The interactions open up new possibilities, such as herbivory among birds like *Jeholornis*, which in turn change the nature of the forest and broaden further potentials. The greater the number of species that live in their place, all with their own evolutionary backstories and traits, the more unique interactions can occur that nudge evolution into varying directions. Diversity creates more diversity, both a result of the ongoing process and also a bulwark against extinction. There are mass extinctions yet to come—an asteroid speeding across space due to strike in about 59 million years from this particular moment—and the greater the spread of different niches, adaptations, and behaviors, the greater the chance that some forms of life will survive the devastation. In fact, it's the beaked, herbivorous birds that will survive what all other dinosaurs cannot. What *Jeholornis* is doing today, merely following an urge to feed itself, is a small action that is helping some birds evolve the traits that will allow the avians of the distant future to subsist on seeds, nuts, and whatever plant material they can find stored in the ground in the days after the Cretaceous world ends in fire.

The deep connection between plants and animals even extends after death. A decade ago, not far from where *Jeholornis* is feasting, a huge sauropod perished alongside a river that slides through the forest. Heavy rains fed a local flood that picked up the decaying *Ruixinia* and dumped it downstream in a bed of sediment over the margins of the river's usual bank. The dinosaur's flesh soon decayed beneath the river sediments, now turning to soil, as plants began to grow in the earth covering the bones. The minerals in those bones didn't go to waste. Plants can use them, too, the roots of young trees and shrubs reaching down to grow over and around the skeleton, etching the bones as they absorb materials from them just as the earliest land plants weathered the rocks that they grew upon. The dinosaur had become fertilizer, nourishing what would eventually become a vibrant patch of riverbank forest.

The trees have their own complicated afterlives. In the same spot, a weakened and dying magnolia was toppled by the force of the floodwater. It's been rotting since then, becoming a little softer and punkier every year. Mammals have sheltered in the hollows of the downed trunk, and plenty of freshwater invertebrates—small crustaceans and snails—have made their homes within the dark, wet tissues of the rotting log. The dead tree provides a home for just as much life as it did when it was alive, just of a different sort. And for our hungry *Jinzhousaurus* looking for something a little more filling, the easily shreddable wood makes for some excellent roughage. Walking on all fours, mitt-like hands leaving dainty impressions in the soggy ground along with those of the broad, three-toed hind feet, the dinosaur ambles away from the shallows and up to the rotting log with its deep tail swaying behind it. The herbivore wastes no time digging in, picking the exposed, fibrous end of the trunk and taking a large bite. Her big teeth make short work of the softened wood, and she soon takes another bite that sends insects and small crabs scurrying for safety among the deeper recesses of the log's crevices. Some of the invertebrates don't make it out. As she munches, the dinosaur inadvertently gets some extra protein from the various little bugs hiding within the decaying ecosystem. The dead plant material and the crunchy little animals will help her maintain her body and tissues until she, too, perishes, perhaps returning those ecological investments to all that will feed upon and grow from her body. The forest remakes itself even as it seems to come undone.

7

A Sticky Situation

100 million years ago
Myanmar

THE TINY BUG NEEDS A REST. ALMOST ANY SPOT WILL DO. Her world is not one of vast seas, towering mountains, or broad deserts, but a damp and gnarled landscape of branches, trunks, and leaves that provide a seemingly infinite number of places to alight. She just needs a few moments before taking off again to follow the carbon dioxide trails of other creatures filled with vital, rushing blood.

Her forays in search of food aren't usually this much of an ordeal. Today, wind has been the problem. Strong breezes have come with another day of drizzle, a chaotic and lilting series of gusts that have required extra effort with every flap of her transparent wings and every maneuver through the vegetation. She cannot hear the sound, but all day the needle-covered branches of the forests' conifers have been creating a soft and prolonged susurration as the storm slowly expends itself. She can't wait for it to be over. For such a small life, food is burned quickly and hunger is a familiar sensation. If she doesn't eat every four days, she'll perish. This is the third day of rain and interfering winds. She can't count on eating tomorrow. The mosquito must try today.

At least these forests are rife with sanguine creatures of all sizes. Some are small enough to see her approach while others might as well be mobile mountains of flesh. There's always a warm body in these Cretaceous woods. Her latest meal, taken on the last sunny day before the unpleasant mist set in, was a broad-headed amphibian clinging to a lofty branch—*Yaksha*. The amphibian had evolved into an ambush-based lifestyle that chameleons would later copy, waiting and watching for unassuming invertebrates to get too close. Pull focus, move slow, and *blap*—the amphibian's spring-loaded tongue can shoot forward to adhere to the creature's unfortunate prey and reel it back toward small jaws set with tiny teeth.

Yaksha might have eaten the little mosquito if she had flown in from a different direction, but she was lucky that day and the little syringe-mouthed insect was able to gently land on the arboreal amphibian's soft skin, plunge her mouth into the pliant flesh, and drink her fill, distending her abdomen with vibrant and translucent red as she heavily flew off to digest.

Her takeoff had to be awkward. There's no dignified way to take off fatted with so much blood that she's three times

heavier than before she started sipping. Every time she fin-
ishes slurping, she begins a frantic flapping of her little wings,
spreading the tiny wisps of her legs across the body of her
unwitting host to hopefully spring into the air undetected. In
a split second she's flying, wings humming at eight hundred
beats a minute to carry the precious fluid inside of her. That
was a good day, and she was nourished enough to lay eggs in
a shallow pond formed in a round dinosaur footprint. Now
she's much lighter, practically empty of the food she needs.
Despite the drops falling everywhere and wind that might
blow her off course, she has to hunt if she's going to survive
through tomorrow. She stretches her jointed legs over the
bark of the tree, as she's done hundreds of times before, and
vaults into the air.

She doesn't get far before an unexpected gust pushes her
back toward the conifer she has just taken off from. And this
time, she's unlucky. The breeze sends her tiny body tilting
and shifting sideways into an oozing, golden mass of resin.
She's stuck.

Dinosaurs can be hard on trees. The biggest ones, incredible long-necked herbivores, scrape against the bark,
snap branches, and even topple entire trunks as they move
through these conifer-filled woodlands. It's just a part of how
they navigate their world, completely unaware of the consequences of each footstep and every scrape against the rough
bark as they pass through the woods. The forest is full of
dinosaur-sized trails that the titans prefer to walk through,
the open spaces further trampled by all the other species that
take these convenient footpaths. But even the most regal and
impressive dinosaur sometimes has an itch to scratch. Not
long before the little mosquito supped on amphibian blood, a

huge sauropod wandered by and rocked back and forth on its column-like limbs, scratching its louse-bitten hide against the rough conifer bark. The biters are too tiny to defend against, tucking into the various cracks and crevices of the dinosaur's scaly skin, and there's little an irritated dinosaur can do but try to alleviate the itch. And in that moment, as the dinosaur scratched its hide against the tree's rough bark, it opened a wound on the conifer. Just as the trees evolved bark to help protect their more vulnerable tissues inside, the dinosaur evolved thick, pointed osteoderms along its flanks—bony armor floating in the skin that could break the tooth of a carnivore that gets too bold. As the dinosaur itched, the spike-like osteoderm scratched over the tree's trunk and caught on the bark, the keratin-covered armor sawing into the outer layers. The tree's response was involuntary. It could feel no pain, but instead the plant's tissues began to respond to the insult. As the dinosaur ambled away, great tail swaying behind its multi-ton body, golden goop began to trickle out of the tree where the wound had just been cut.

An open gash leaves a tree vulnerable. The dinosaur had unknowingly exposed the inside of the tree to fungus, wood-chewing insects, and other hazards that could take hold in the wood and slowly make the tree's own body hostile to itself. Even a relatively superficial opening could be an entry point to living things that would consume the tree from the inside out. Resin evolved as a kind of natural bandage, able to cover over and harden until the tree's slow growth could replace the damaged tissue with a new protective layer. Following the scratch, the tree's resin dripped, oozed, and sealed the wound, longer threads of the golden liquid trailing down the bark and a few dropping onto the forest floor. At least the ooze was performing as intended. Sometimes the resin can build up among the cells between the wood and the bark, and if enough of the sticky goo accumulates, the tree trunk can burst from the built-up pressure.

Such catastrophic malfunctions aside, most of the time conifer resin slops over the gash and remains tacky for days afterward. Creatures climbing over the trunk and even debris carried by the wind can get stuck in the natural trap. Where conifers grow along the coasts, even seashells sometimes get blown into the goop.

Amber does not perfectly encapsulate ancient life, but it is a unique form of preservation that helps transform once-living organisms into fossils.

The sticky stuff that had suppurated from the tree's bark will coagulate and solidify in time, but not soon enough for the mosquito. She was blown right into the center of it, too small to pull herself free from the mess. She struggles. She has to. She has to try to pull herself free with every twitch and flutter her little body can muster. She can't feed and she can't drink stuck to the tree. But for something so small, the resin is fatal. The thrashing of her final moments become recorded in the resin itself, her wings outstretched among tiny bubbles of oxygen stirred into the yellow mire. In a few moments, she's encased. She can't breathe, the golden liquid a barrier between her and the outside world that deprives her of the vital oxygen the mosquito needs. She is part of the tree now.

The same story is playing out all over the world, completely mundane moments of an exceptional time. In woodlands much like this one in prehistoric Myanmar, from equatorial forests to those that survive through the long dormancy of polar nights, pieces of Cretaceous life are becoming encased in resin. A gash in a pine near a beach suppurates golden glop that catches the vacated shell of an ammonite kicked up by a gust blowing over the sand. Chewed-up feathers of a nesting toothy bird become stuck in the fresh resin of the tree hosting the nest, trapping the larvae of feather-eating beetles along with the decomposing plumage. A big-eyed lizard takes a few unwary steps up a tree trunk and becomes trapped in a sticky puddle, soon covered over and encased. This is the Cretaceous Resinous Interval.

Plants capable of secreting amber-producing resin have been around since the Carboniferous and the days of the scale trees. Until now, such trees have been relatively rare and often exuded resin from their roots rather than along the branches. At least a few dinosaurs have scraped their feet on the forest floor because they accidentally stepped in a sticky patch. Such trees typically didn't trap insects and other signs of life, however. What's happening now is different, a special time in plant history that is almost entirely reliant on the greenhouse climate of the Cretaceous world.

Earth's resinous interval isn't a distinct geologic time period itself but more a phenomenon, a consequence of a global climate that's favored dense forests of resin-producing trees. Even though angiosperms have been around for more than 25 million years at this point, they are not nearly as diverse, disparate, or widespread as the gymnosperms that have formed the core of primordial forests for over 150 million years. The world's greenery is still mostly made up of cycads, rough and stubby bennettitales, and, most of all, the brushy-branched podocarps and other pine relatives. Eventually, after a catastrophic extinction, angiosperms will have their time and some will become capable of oozing amber-producing resin, too. But this is the heyday of the conifers that are collectively pouring out gallons and gallons of resin in forests the world over.

The Cretaceous world is a warm and humid one, kept sweltering by outgassing of undersea volcanoes pushing the Earth's seams apart by bleeding new, molten rock. The carbon dioxide and methane released from deep within the planet have fed a new warming pulse, another chapter in the seemingly endless summer of the dinosaurs. It's hot enough that ice at the poles is a relative rarity; forests are able to expand throughout each continental landmass closer to the poles than at almost any other time in Earth's history so far.

Even the arrangement of the continents themselves has a role to play. Much earlier in the Mesozoic, in the time of the supercontinent Pangaea, the continental interior sometimes became dry and covered in vast, dune-dotted deserts. Habitats closer to the poles experienced seasons and could be chillier, but the continental interior was largely shielded from the cooling effects of ocean currents and other abiotic forces, becoming incredibly harsh and hot places to adapt to. The interior of the landmass was too arid and warm for expansive forests to take hold and so dense groves grew out toward the edges. As the continents were pushed around into new positions, currents in the seas and the air altered, shifting weather patterns as new coastlines opened in places that had once been deep in the interior. By this Cretaceous moment, the average global temperature sits at about ninety-five degrees Fahrenheit and resin-producing trees have thrived in the global hothouse. The lavishly branching *Araucariaceae* trees, the monkey puzzles, have done especially well—and are among the most enthusiastic resin-producers of all their coniferous family. They've been around for nearly 100 million years, providing fodder for all sorts of sauropods and insects and other forms of Mesozoic life, but it is in the Cretaceous that they spread far and wide to an extent never seen before. By this point, they had been oozing copious amount of resin in the world's forests for 25 million years and will continue doing so for at least another 25 million more.

The incredible quantities of resin are creating a new fossil record, one that has greater potential to enclose soft tissues and tiny creatures that are often too delicate to have any hope of being preserved in coarse grains of sand. Now resin is providing the initial step, enfolding organisms that live on and among the trees that would otherwise tumble to the ground or decay within the creases of rough conifer bark. From there, the hardened resin has to find its way into the rock record, the yellow drips giving once-living things a better chance of surviving the transformation into a fossil.

For a few days, the mosquito is little more than an un-noticed speck in the forest. The resin surrounding her body becomes slicker and more pliable in the heat of the day, hardening again each night as temperatures drop. Her compound eyes behind those layers of resin can't detect the shifting sunlight or the fluttering birds that chatter and chase insects through this part of the forest. But the next dinosaur to happen across the same tree makes the itchy sauropod seem absolutely gentle. The conifer's branches look too good to resist to the next sauropod that wanders by, the monkey puzzle shimmering with dew scattered across innumerable needles in the early morning light. Despite being bigger than anything else in these groves, the rust-and-cream–colored dinosaur is far from being fully grown and can't quite reach the tastiest-looking branches while on all fours. So he rears back, bracing the jellybean-shaped pad of his front foot on the tree trunk as he sits on the thick base of his bracing tail and reaches his neck upward to take a big, spoon-toothed bite of the greenery. He can't stay in this position for long—his massive heart has to put some extra squeeze behind every pump in order to keep sending blood so high up into the air—but the tree doesn't give him much of an option. Nor is the tree capable of holding up so many tons of dinosaur. He's too massive and the tree gives way, toppling below his bracing foot as he trumpets in alarm and small pterosaurs screech as they flutter off to a quieter part of the forest. Still, he has the good fortune to land on his feet, and at least now the branches are at a much more amenable level. The side of the tree containing the resin-ensconced mosquito is pushed down into the soft mud as he nips, grazes, and swallows, taking his fill before moving from tree to tree to satisfy the almost ever-present feeling of emptiness in his great belly.

Uprooted, and stripped of its photosynthesizing needles, the pine soon dies. Birds pick at the exposed bark for grubs. Moss grows over the cragged surface and the wood slowly becomes punky as it rots through the seasons. The pine is now in its afterlife, a home for different forms of life just as the decomposing body of a marine reptile can become a reef. The hardened wood that reached dozens of feet into the air starts to become soft, duck-billed herbivores contributing to the disassembly by munching on the rotting wood and the various fungi, mollusks, and crustaceans within. But no one wants the old resin, cracked and detached from the tree. It doesn't take much flow for a heavy rain to collect the piece from the decaying log, and many others this patch of coniferous forest has produced, and wash them together in a small stream, tumbling and mixing each dollop with sediment that buries the mosquito's tragic end. Now she has truly begun her journey toward becoming a fossil. As the rest of the tree is broken apart and its molecules distributed through the ecosystem, season after season, the mosquito is being welcomed into the fossil record.

The process doesn't happen all at once. The mosquito has been trapped, covered, washed away, and now buried, representing a fleeting moment of Cretaceous time that is already long past. The fate of the unfortunate little bloodsucker might have been very different had she become stuck in mud or perished by some other means and tumbled, unnoticed, onto a stream bank. The pathway through which fossils become fossils is unique from one organism to the next, a series of filters that operate through luck and happenstance to lock a piece of the past into stone.

Despite the fact they live in incredible numbers, forming an essential part of the world's ecology, most insects will never make it into the fossil record. If a tiny insect doesn't become prey for some other creature—and Earth's ecosystems are endlessly hungry for insects—she would be so small that only the most exceptional circumstances could have blanketed her lilliputian body. Something as ephemeral as an ash shower from a local volcano blanketing the landscape with flecks of soft sediment would have some chance of burying something as delicate as a mosquito, but otherwise even grains of sand might be too coarse to safely enclose her body. Conditions have to be just right, the mud or clay just fine enough, to tightly wrap around the tiny insects that so greatly influence Earth's ecologies that even the great dinosaurs wouldn't exist without them. The fact that she happened to be born in a grove full of resin-producing trees, roamed by dinosaurs that had taken to opening dendrological wounds as they scratched themselves, gave her a better chance at becoming a fossil herself. Now she's ready to undergo a great transformation.

When a once-living thing is covered over in resin, it is not flash-frozen or smuggled into the rock record unaltered. After all, even the resin has a transformation to undergo. Amber is altered resin, not the original goop itself, just as the sandstone encasing so many dinosaur bones starts off as a chaotic association of grains and granules that eventually become transformed into something else. Change is the very nature of the fossil record, transformation into something different—a riff on what once was—critical to an afterlife that can persist for millions of years.

The tiny mosquito is still changing even years after her death. She's buried beneath part of a stream running through the woods she called home. Sediment carried in the water begins to settle on the layers above where she's buried, creating more weight and pressure on the collection of amber pieces below. Resin is incredibly resilient, but it has its limits. Beneath the ground, the resin begins to crack along weak points. It's these tiny breaks that will help her withstand the test of time. As groundwater continues to seep through the stream layers, it seeps into the sanctum of the resin and begins to soak into the body of the long-dead mosquito. That water carries various minerals that had dissolved out of the rocks and soils in the area, silica, calcium carbonate, quartz, and more, chemical compounds eager to bond with the organic tissues of the mosquito's body. The minerals infiltrate and replace her tissues, beginning to copy them and even fill in the tiny cavities within them. Her body is acting as a mold for what's becoming a mineralized cast. Little by little the mosquito is being assimilated into the fossil record, copied down to her most delicate structures—a facsimile that will let the world know that she was once here, part of our planet, when the golden drop of amber one day sees the sunlight millions of years from now.

8

Rainforests and Revival

60 million years ago
Colombia

THE SHADOW MOVES. IN SLOW, SWEEPING UNDULATIONS, the forty-foot apparition slides beneath the weed-choked surface turned to gray glass by cover of clouds above. It's not a trick of the light. The sinuous void is alive, an adult *Titanoboa* searching for a quiet place to rest on another muggy neotropical day.

Earth hasn't seen reptiles her size in quite some time. Even though she's not quite so bulky, owing to a streamlined body shape that slides and slithers through the swamps, the last time forty-foot reptiles wandered the planet was about 6 million years before the present moment, before an asteroid changed the course of life on Earth forever after. Much of what survived the days of fire and years of chill were small. The creatures that were able to get underground on the first day and survive on little during the chokehold of impact winter had the best chances of survival. Now, in these early Paleocene days, descendants of those survivors are beginning to evolve and intertwine in new ways—and some animals are starting to live large once again. *Titanoboa* is among the new giants, although her impressive girth is not a signal that the Age of Reptiles is about to get another chapter. Life is not merely picking up where it left off before its fiery interruption. Earth's ecosystems were so thoroughly razed that whatever survived is building something entirely new. Her world is one that would be unfamiliar to *Tyrannosaurus* and *Triceratops*, a dense and sweltering grove where the world's first tropical rainforests have enthusiastically taken root.

For this particular place and time in the Paleocene world, the day is not especially hot. The air temperature hovers around eighty-seven degrees Fahrenheit. And despite the ominous clouds stretching from horizon to horizon, the morning is more humid than it is actively wet. Only a few drops ripple the surface of the broad, sluggish river as *Titanoboa* slides toward a suitable resting spot to while away the afternoon, now and then flicking her dark tongue above the surface in case scent betrays a possible meal wandering close to her hiding spot.

It's really a perfect environment for an ambush predator. A large terrestrial hunter would struggle to make their way through the dense groves of trees without pushing them over, and even these patches of relatively dry land are small islands between rivers, ponds, swamps, and stretches of slippery, sodden ground. A carnivore like *Tyrannosaurus* would be sliding and swimming as it tried to stalk prey, struggling to keep their footing among the waterlogged woods. *Titanoboa* is better suited to the soaked landscape. She can swim beneath the greenery growing thick on the surface of the water, slide over the muddy embankments, and, like the crocodiles she lives among, hide in the shallows, resting until the unwary forget what lurks in the water. An abundance of fish keep her well-fed, but she's not one to pass up clumsy mammals who wander too far from the comfort of the tall forests that have sprung up so enthusiastically that no space on the forest floor is wasted. Gripping with her sharp teeth, she wraps her body around the beasts and hugs the life out of them.

The mammals that chatter and clamber among the trees above the snake's head are living in forests unlike any Earth has seen before. The change stems not only from the plants that survived the catastrophe of 6 million years ago, but the constant push and pull of animals and plants living together.

On the last day of the Cretaceous, when hadrosaurs honked and *Triceratops* shoved each other with their great three-horned faces, Earth's forests were mostly made up of conifers. Trees like magnolias and broad-leafed palms that shaded the scaly backs of enormous crocodiles had been around for millions of years, but their spread was always a push for space against the ever-present gymnosperms. Angiosperms spread in vacancies that toppled monkey puzzles and pines left open, assisted by seed-dispersing birds, mammals, and pterosaurs. Big dinosaurs sometimes assisted with the slow shift, too, even though they had no idea they were doing so.

Large dinosaurs pushed over trees, trampled down vegetation along game trails, busted up rotting logs, and munched on what seemed to be a never-ending buffet of verdant fiber. Forests had to evolve around their habits, meaning that most had open canopies that let plenty of sunlight down onto the dinosaur-sized spaces between the rough trunks. Trees didn't knit together into a big, shady canopy over the ground but often kept each trunk relatively isolated from those surrounding it. Small moments of connection—a squirrel-like multituberculate running along the branch of one tree to its neighbor or worker ants crawling from one tree to another in search of food—were relatively rare happenstances among these dispersed stands.

The extinction threw everything into chaos. About 45 percent of plant species perished, never to return. Connections between plants and the creatures that relied upon them were also severed. Various insects that ate leaves, bored into them to settle their larvae, and otherwise treated plants as both food and shelter went extinct, too. Their hosts—and their homes—were destroyed by fire and suppressed by three years of reduced sunlight that almost halted photosynthesis the world over. Locales closest to the impact site in ancient Central America were hit hardest, coping with the earthquakes, tsunamis, and other after-impact effects that other parts of the world were spared, but there was no place on Earth that was untouched by the consequences of the asteroid's unexpected smack. The impact created an intense moment of rapid, violent, churning death. And yet not all was wiped away. More than half of plant species survived into those early, scorched and chilly days of the Paleocene. There could be no predicting what would happen next. The story of recovery would be narrated in the spaces where each life met others, the interactions that would create a new community from what had been spared and had survived.

Earth's regreening didn't happen the moment the debris clouds dispersed and sunlight warmed the ground. Plants didn't recover their pre-extinction diversity levels for about 6 million years. The vegetal survivors competed with each other for spaces to grow, naturally, but many also developed new relationships with the surviving insects, birds, and mammals that became their pollinators. What began as homogenous mixes of survivors began to differ, finding varying evolutionary inflections such that more species could thrive in the same place. The aftermath of the impact even gave plants a little assist. Ash from widespread wildfires as well as intense volcanic activity among the Deccan Traps in ancient India settled into soils all over the world. The flaky debris was rich in phosphorus, a critical mineral that plants use in everything from photosynthesis to transporting nutrients through their bodies. Early bean plants and other legumes began to spread, and consequently, these plants brought more nitrogen into soils that soon became more fertile and capable of hosting forests, where individual shoots raced each other for access to sunlight. Forests that were left as nothing more than smoking ash on the first day of the Paleocene went through a succession of stages, carpets of ferns giving way to multitiered woodlands. And plants didn't simply stop there, as if reattaining a prior species count was a goal or even something they could be aware of. Following shifts to the Earth's climate, plants thrived as never before.

The warmth of the Paleocene world led to vast amounts of methane to be thawed and released from the deep oceans, driving a warm spike that all but eradicated global ice. The time of *Titanoboa* is such a moment, a time when even polar locales are closer to sweltering swamps than ice caps. Elsewhere on the planet, within the Arctic Circle where darkness makes itself at home for months at a time, Earth's northern pole is covered in swampy groves of the conifer *Metasequoia* and where the hippo-like mammal *Coryphodon* sloppily smacks up water plants and crocodilians lazily swish their scaly tails through the water. It's a great heat spike before a gradual cooldown that will keep refashioning Earth as it goes. Over time, the greenhouse gases in the atmosphere will dwindle and distinct seasons will return, rain patters and currents shifting with the climate. For now, though, the world is warm and plants thrive from pole to pole, creating new forests that are the crucible for the creatures that will spread over the land through the next 60 million years.

Within Colombia's ancient Cerrejón woods, it's angiosperms that are now most vibrant. More than fifty different angiosperm species grow together here, the once-widespread ferns and conifers making up only a handful of holdover species. And these are plants that are well suited to the warm and the wet. The landscape receives over 120 inches of rain each year, so much so that the plants have had to evolve ways to avoid becoming oversaturated. Here, leaves are broad and bear smooth margins that taper into long points—drip tips. It's such a botanical paradise, in other words, that plants have evolved unique structures to help excess water run off their leaves so the water doesn't collect and start growing algae that would steal the sunshine of the larger plant. Every leaf is important for survival. And the angiosperms don't just grow next to each other in discrete and orderly fashion. Long lianas drape themselves among the branches of broad-leafed trees related to laurels and willows, whose reflection ripples in streams where lily pads huddle close together and inadvertently conceal the giant snakes and sleek crocodiles that dwell here. Early arums, teas, and peppers grow along the stream banks where they can sneak in some sunlight away from the thick, closed-up canopy of the forests.

The botanical variety here is incredible, species growing over and upon each other in a virtual curtain of greenery. The very sunlight that plants have depended upon ever since their origin, however, is not unlimited. Every leaf, branch, and trunk casts shadow, and the more the canopy snarls around itself, the less light reaches the forest floor. The success of some plants deprives others. Sown into uncertain ground, every seed is at the mercy of both the elements and where the nascent plant's neighbors cast shade. Every one of them is reaching toward the light.

Growing along the margin of the water, a grove of palms has managed to peek out from under the shade of the canopy and grow tall. From a thick cluster of stems, the palm leaves stretch up to thirty feet in the air. It looks like a collection of bright-green feather dusters huddled together. They follow the slow seasonal path of the sun. Plants have had to do so from the time of the earliest forests, the anatomical structure of the greenery allowing them to tell the best way to grow.

The way the rainforest's palms and other plants detect the most advantageous angle for more sunlight plays out on the microscopic scale. The cells of the swaying palms are filled with water and have cell walls to help seal that water in. Between the cell spaces, like an air ventilation system, are tiny airways that are relatively dry. What light does as it moves between the cells and the air spaces, then, allows the palms and other plants to sense where sunshine is coming from and grow toward that source even as they remain firmly rooted. Light, after all, has its own behaviors. When sunlight moves between the wet cells and dry air spaces, it scatters and changes direction. With so many cells and airways, the repeated refraction is like a collection of data points that directs plants to where the essential sunlight is coming from. It's not the specific green of the plant, but the plumbing that allows plants to follow the sun as the Earth's orbit and rotation change where light falls.

Even the chloroplasts inside the leaves move to follow the sun's repeated arc. As plants grow toward light and shift their leaves for maximum exposure, where light and shade falls upon the plant changes. Leaves that more fully shift their broad expanses to the sun can throw other leaves below or behind them into shade, the plant getting in its own way. To compensate, the chloroplasts inside the plant cells move. Especially in shade leaves, or those that usually don't get the best access to sunlight, chloroplasts can jiggle into a better position to make the most of whatever sunlight hits the leaf. As leaves shift to face the sun, even the interiors of their cells change to make the most of every photon traveling across space to touch the face of the plant.

Making the most of every sunny day is essential, especially as plants have never been left alone to grow unbothered by the world around them. From the time plants first started to grow on shore, there have been living things waiting to nibble on them. The explosive abundance of green food has not been ignored by the herbivores. Some plant-munching mammals are becoming better suited to mashing leaves and fruit to pulp, opening evolutionary routes that will eventually lead to hulking, hairy beasts with fermenting vat stomachs and impressive, tree-destroying molars. For now, though, the mammals of the early Paleocene are little different from their relatively small Cretaceous predecessors. Some have become larger in overall body size, an abundance of food and a lack of dinosaurs allowing the mammals to push body size boundaries previously impossible for them, but they're ballooning in size so fast that their brains are still the same size as those of their ancestors 6 million years ago. It's going to take millions of years more of everyday interactions for novel behaviors to evolve, dynamics within and between species that will require more complex ways of remembering and thinking. In this moment, the surviving mammals are more or less supersized versions of their Cretaceous ancestors that have yet to truly undergo their big evolutionary burst. No, among the herbivores, it's the insects that are more important.

Of all animals, no group has influenced plant evolution quite as deeply as insects. Even during the dinosaurian heyday, the big reptiles shaped the landscape and the nature of the forests, but insects have had a longer-standing and more intimate relationship with plants. Insects live among plants, munch on their leaves for their food, drill into their bark, lay their eggs upon and within plants, and reproduce so rapidly and in such numbers that the tiny, almost inconspicuous creatures have an outsize effect on what plants have and will become. It was insects, after all, that became some of the first pollinators tens of millions of years before these Paleocene days, picking up pollen by accident and transporting it to different plants as they foraged about. While it's possible to get a rough sense of how an environment has fared by looking at the charismatic creatures—the giants like *Titanoboa*, the carnivores that imply the existence of abundant prey that in turn need plenty of food in the ever-shifting array of ecological connections—the fine detail is better seen among some of the smaller inhabitants of these tropical forests. It's a way of looking at an environment's foundation, the base-level interactions from which so much else flows.

The Cerrejón is not a pristine garden. A grove flush with perfect, undamaged leaves would be a sure sign of something gone terribly wrong, plants grown in isolation as if under glass. Every other leaf here, half of the forest from the edges of the water to the top of the overlapping canopy, shows signs of having been nibbled, chewed, bored, and otherwise familiarized with the ever-busy mouthparts of countless insects. It's what makes this a thriving forest, one in which life is constantly traded and exchanged through all its wonderful and varied permutations. Life is messy, ragged at the edges, and these tree stands surely show signs of being lived in. The partially chewed state is a return to arthropod influence at the end of the Cretaceous, although the cast of characters has changed.

A large part of insect success is that they reproduce so rapidly and in such numbers that potentially useful variations show up quickly and can replace what had been lost. The result is a great moment of convergence—the hardy survivors of the end-Cretaceous cataclysm formed a base from which beetles, caterpillars, and so many other ever-hungry arthropods could reinvent roles that had been temporarily erased. From chewing on the margins of leaves to rolling dung and burrowing through bones, different insects have carried out the same roles over the course of evolutionary time, their forms shaped by the opportunities presented by the other organisms they live among. Of course, the forest is not populated by giving trees that peacefully acquiesce to the nonstop mining and munching. Constrained in their movement, the plants of this forest can't simply move away from the biting bugs like a giant snake being bothered by hungry mosquitos. The plants' strategy is more like tower defense, employing methods that make them more bothersome to turn into fuel in invertebrate guts.

Plants have evolved defenses almost as diverse as the number of genera. Some load their tissues with harsh compounds that are either noxious to insects or intoxicate them, encouraging the insects to go bother another plant or at least slowing down their munching. Structural defenses like thick coats of wax on leaves may prevent some insect species from penetrating the plant tissues altogether. Nor are individual plants entirely on their own while under invertebrate assault. Plants can release chemical compounds that let their same-species neighbors know that an attack is likely coming, spurring those nearby plants to start building up repellant or defensive compounds in their tissues. Some of these cues even attract insect predators, a call for an assist that benefits both plant and its bug-munching collaborators. The back-and-forth is not a matter of natural balance, but more like a drawn-out evolutionary conversation. New plant defenses unintentionally select for insects with ways to get around them, which in turn help bring about more resilient and resistant plant species. New variations, shading into entire new forms and adaptations, maintain the stalemate.

Paleocene forests were a crucible of evolution for mammals,
providing dense habitats for new beasts to evolve among.

High up in the canopy, so deep within the spread leaves
that the small beast can't even see the ground, a squirrel-
sized multituberculate scampers along a thick liana. Each
reach forward with its small, clawed paws gives the mammal
enough stability to kick its hind feet forward, a cycle of reach
and push best suited to a life that rarely ever touches the
ground below. He doesn't stop until he reaches the relative
stability of an extended branch the liana dips below, a firm
place to sit back, flick his poofy tail, and chatter a rolling series of clicks that reestablishes—for the twentieth time this
morning—that he doesn't desire any company in this patch
of the Paleocene forest.

Not all multis are so small. Thousands of miles away, in
what will one day be the vicinity outside Denver, Colorado,
there shuffles a multituberculate about the size of a golden
retriever, *Taeniolabis*. The mammal's chisel-like teeth and
many-cusped molars associate it with the treetop scurrier in
prehistoric Colombia. They are among the most successful
mammal lineage that will ever exist on the planet, beasts that

are more than ready to take advantage of a world that is the same size as it ever was but somehow contains even more space to thrive.

The first multis started scrabbling over tree bark more than 100 million years before *Titanoboa* slithered through the swamp. They were among the first mammals to care for their fuzzy, pink babies, lavishing them with milk so that they could grow quickly. A little bit of parental care went a long way—multis spread across the ancient Earth, munching on ginkgo fruits, gnawing on the bones of dinosaurs, and even creating their cozy, warm burrows underground. The long lead-up, all that time to stumble into different ways of being, allowed some to survive the terrors of the impact. The multis that were able to hide even a few inches underground survived, and their multipurpose teeth—capable of chawing bark as well as seeds and singed dinosaur bone—let them chew their way to the salad days of this humid forest. In the same acre of forest that once grew where the territorial little multi belts his tune, there is now more habitat in the same place. It's forests like these that will be critical to how mammal evolution flows from this point onward.

In the years soon after impact, the number of ways a mammal could interact with the environment was limited. Closed forest canopies were rare and so mammals mostly lived along the surface of the ground or beneath it. A burrow was a great shelter from many things, be it storms, the heat of midday, or hungry dinosaurs. Beasts were tightly tied to the soil, their lives contained within a few inches above and below. As trees began to reach higher and higher, some mammals could begin climbing the trunks, foraging and even making homes within hollows. Better climbers passed on their subtle skills to their offspring, repeated over and over again into increasingly arboreal mammals. The constant search for food and shelter on the ground, too, began to shunt mammals in different directions. Many were still generalists, but some began to favor low-growing plants while others plucked insects from the stems and leaves. Both sorts could be prey for those with slightly sharper teeth. Still, this world of open woodlands wasn't yet all that different from the one the dinosaurs had inadvertently created. It wasn't until more than a million years after impact that all the dendrological buildup began to create something new, forests with possibilities never open before. Mammals could evolve natural histories that never need touch the ground, lives spent entirely between the trunks and branches of trees that practically embraced each other. The closed canopy altered plant life below, as well, removing much of the low ground cover into a more open understory where predators like the mesonychids, "wolves with hooves," could stalk their next meals. Burrowers can still scrape away into the soil, climbers finding opportunities for forage and escape among the rough bark of so many high-reaching trunks, the forest creating the equivalent of a multi-level apartment instead of a single-level with a finished basement. More life can thrive here, the slightest changes to light, moisture, and the spread of levels opening potentials that would not only allow mammals to thrive, but to begin living large.

9

Adrift

40 million years ago
The Southern Atlantic Ocean

THE ENDLESS SEA FEELS CALM, LIKE A SLOWLY UNDULATING pane of blue glass even as the current still moves all that floats among the salty expanse. The spine of the night is just beginning to arch overhead, somewhat muted against the lunar glow but visible against the dark. And out in the middle of the seemingly endless and ceaselessly shifting water, there is a tree.

The tree is miles from any island, or even any continent. It's been floating for days now, directed only by the flow of oceanic water and breezes that catch in its botanical sail of leaves. A trio of tiny monkeys huddles close to the trunk, tails twined with one another's beneath the gently rocking branch. They didn't choose to take this journey across an entire ocean. The weather hadn't given them much of a choice.

The small primates aren't the first of their evolutionary family. The very first primates were twitchy-snouted insect eaters that ran through the trees above the head of *T. rex*, survivors of the world's worst catastrophe that persisted into a world that was soon covered in lush forests that could cradle the small beasts through evolutionary time. Living in the trees went from a way to largely avoid the terrible lizards to a haven for the small beasts. While the first primates were almost like shrews, these dawn primates quickly began to evolve new traits that would differentiate them from the squirrel-like multituberculates and other mammals that shared their arboreal habitats. Forward-facing eyes granted primates binocular vision, the ability to judge the distance to the next branch or that tasty-looking beetle crawling over a leaf. Grasping hands, rather than claws, allowed primates to both traverse the stems and branches of the trees and handle their food, snatching fatty moths from the air and plucking fruit from their stems. Primates did not just scramble over the tops of tree trunks but grabbed on to their dendrological homes and held on tight, a relationship so intimate that they became shaped by the forest.

The primates huddling close among the sodden tree branches are closer to being monkeys than many of their predecessors. Millions of years from now they'll be labeled as eosimiids, small and fuzzy primates with large eyes and grasping hands who often dine on insects, fruit, and other snacks filled with fat and sugar. Their realm was the coastal forests of pre-historic Africa, not an improvisational island drifting across the sea. But during this age when intense seasonal storms rip along the African coast with each monsoon season, being so tightly tethered to the trees came with an unexpected conse-quence.

When they last saw their old home, it had been the wet season. Some days were merely overcast, coating everything in a fine mist. Even on days when it didn't rain, everything seemed to drip and carry a wet sheen to it. The patter of drop-lets falling from one leaf to another filled the daytime forest. But there were storms, too. The muted grays overhead could quickly turn menacingly dark, wind gusts that made the for-est thrash and dance an early warning of the fat drops of rain that would soon begin pounding the floodplain. Places to hide from the monsoon were limited for the half-pound primates. Especially while out foraging, far from the cozy tree holes they slumbered in every night, it was easy for the little monkeys to be caught among the branches, holding on with their tiny hands, feet, and tails while leaves lashed and heavy fruits bobbed around them. Sometimes all you can do is hold on.

Many of the storms would exhaust themselves and move on with little direct change to the landscape. Streams and rivers might run swollen from the downpours, spilling en-riched water farther across the floodplain to refresh the land-scape, but their intensity was only temporary. Life would resume once the torrents turned into gentle drips tipping off the ends of waxy leaves. Even so, each rush of sediment-laden water would carve away just a little bit more from the streams and rivers. The rushes carved away sediment from the waterways and undermined the banks beneath groves of coastal trees. The forest would seem stable until suddenly it was not. Against the surges, some parts of the bank would be torn away and float downriver toward the sea. These were not logs wrenched out of the ground or thrown into an organic slurry. Vast entangled chunks of shore would cling to one another, held together by twined roots as they were torn away. Individual trees and even broad sections of forest began to bob along the elevated water levels, skirting over the usual river obstacles that might break them up or snarl them, thanks to the high water. What was once dry land suddenly became rafts, often carrying unwitting passengers.

It had all happened so fast. Two of the primate sisters had been foraging together, pawing and chittering at each other as they searched for the juiciest fruit through the landscape of leaves as a male from another part of the forest sat and groomed on the other side of the trunk. Just then, thunder rolled and the winds began to make every leaf begin to twist and flutter. The monkey-like mammals retreated closer to the trunk along larger branches that seemed more stable and less likely to give way against all the swaying and shaking. But it was not the tree that would betray them. The storm surge scoured the already-loose soil beneath their refuge, causing it and everything its roots were entangled with to slide down the bank like a photosynthetic crocodile doing a belly slide. The monkeys could feel the tree move as it was carried along by the swollen and sediment-laden waters through the estuary and toward the sea, the movement seeming impossible for what had for so long seemed still and stable.

In the wake of such a calamity, the primates are very lucky. The intense storm gave way to relatively calm seas and skies dotted with white, fluffy clouds. A storm out at sea could create waves tall enough to swamp and sink their floating shred of continent. Thus far, they've been spared such a fatal dunking. And there's food. The tree that they desperately cling to, as well as some torn branches from neighboring plants, had been laden with fruit and populated by various crunchy insects when it all slipped into the water. There's enough for the three of them to nibble and survive—not just the flesh of fruit and arthropods, but liquids that will keep them hydrated in an expanse where the only other water is painfully salty to even taste. It's hardly an ideal cruise, but they're already luckier than other primates who have been unwillingly sent on this same journey and never saw dry land again.

The repetitive soft splashes and slaps are broken by a sneeze in the night, not far from the raft. The male primate squeals in alarm, grasping the trunk of the tree, while the sisters bare their tiny canines and bounce along the branches, trying to warn off whatever's out there. It could be trouble. There are large and curious carnivores in these waters, some of the earliest whales. The land-dwelling mammals started becoming more and more accustomed to the water over 15 million years before this moment, paddling through estuaries, chasing fish, and snagging the occasional unwary creature from the banks. Mammals were reinventing what crocodiles had done time and again. Then the whales kept going. Some of the whales with legs became evermore at home among the waves, generations of life in the water leading to the origin of sinuous beasts with wicked smiles best suited to grabbing and slicing their fleshy food. Some of these whales, the basilosaurids, are relatively small and sleek, nipping after fish in coastal mangroves. Others are colossal bottom-grubbers

that built their two-hundred-ton bulk on shellfish and crustaceans hiding in the silt and sand. But the most formidable
are open-ocean dwellers that could stretch more than fifty
feet in length, propelling themselves with up-and-down undulations of their flukes as they pursue sharks and smaller
whales. A curious *Basilosaurus* could easily sink a flimsy raft
of salt-crusted primates.

Manatees and dugongs evolved from land-dwelling ancestors,
just like whales.

The mammals are fortunate. The spout was not from a
hypercarnivorous whale, but a mammal from a different lineage that's been undertaking an aquatic transformation all
its own. It's a chubby sirenian, a relative of manatees and dugongs. The monkeys don't know this, of course. They continue to chatter and cling as the blubbery gray form paddles
a little closer to the raft, curious. The beast doesn't yet have
the broad tail fluke its later relatives will use to slowly swim.
It's more of a nearshore herbivore, pushing and paddling with
stumpy limbs tipped with broad nails at the end, not too

different from an elephant's foot. In fact, the manatee and the prehistoric elephants thriving back in Africa are close relatives, sharing a number of traits in common like incisor teeth transformed to tusks and mammary glands on the chest instead of along the belly. And like some early elephants, this manatee also has cheeks full of grinding teeth that pulverize vegetation.

Not one to let a potentially new delicacy slip by, the manatee opens its whiskery, bristly jaws a moment and munches into the side of the raft. The entire float bobs and shakes as the herbivore tears away a chunk, the cries of the monkeys impossible for the manatee to hear above the encompassing lapping of the surface water. But while the greens looked good, they taste strange—not like anything the seagoing beast has savored before, and certainly not like the seagrass that fuels the sirenian's roly-poly body. She lets the remainder of the chunk she pulled away fall out of her mouth as she snorts once more, kicking her limbs beneath her as she heads toward the west, the same direction that the raft is heading.

It won't be much longer now. From such a moment of torrential terror, the raft has nearly survived its journey across the Atlantic. The island of South America is just about fifteen miles away, beyond the horizon and the ability of the monkeys to see, but there all the same. It will be their final destination, a place where the primates will dry out their salt-crusted fur and begin new lives in an unfamiliar place. An accidental trip will introduce new forms of life to a distant continent.

Currents and tides, wind and vast stretches of water, all seem to be barriers to connection between landmasses. For many living things, they truly are. A saber-toothed, cat-like carnivore that lives in North America at this time has no route to get to South America, for example, unable to swim the distance between the continents and living too far inland to have any hope of being swept off. But ancient South America is not a fortress that repels all who visit.

Islands are not isolated laboratories of evolution. If that were true, whatever creatures happened to be on the landmass when it split from its siblings would only roll through variations of species present during the continental divorce. Distance doesn't require inaccessibility, as the monkeys will soon demonstrate when the crash of a shoreline wave throws them onto the sands of this unfamiliar environment. They lived in a habitat prone to casting rafts of vegetation into the sea at a time of intense coastal storms and a relatively short distance between Africa and South America, around 620 miles, and the weather was with them for this particular journey. They won't be the only animals to take the trip, either. At least two other primate families will be cast off on similar trips, one of which will be the ancestor of the capuchins, spider monkeys, and other primates that spread through the ancient Americas. Without the seabound sweepstakes, South America wouldn't have any monkeys at all. Rodents will join them. The tiny, fuzzy ancestors of the capybara and half-ton rodents that will chew on water plants among the swamps will come from Africa, similarly cast adrift from the same coasts—sometimes on the same rafts—and shoved out onto beaches facing east rather than west.

Rafting to another island or continent isn't a trip any prehistoric organism plans to take, and yet such journeys have connected parts of the world that would otherwise have remained isolated for millions of years. Throughout time, small species such as various lizards have been able to inadvertently island-hop by clinging to storm-ravaged vegetation ripped from one shoreline and pushed onto another. Birds, too, can fly for long distances, transporting parasites, seeds, and small organisms they pick up in their feathers and on their feet in one place and drop them wherever they land, ferrying life to new locales. Even seeds can travel from place to place. Since the time of *T. rex*, at least, Southern Hemisphere plants called screwpines have been casting their multi-faced seeds into the oceans. From the tufts of long, trailing leaves, fruit made up of multiple parts—phalanges—grow and eventually drop, those that make it to the water buoyant enough to float. If the cluster of seeds is lucky, the foundation of the next generation will survive all that salt and become planted along the shore-line of a different island or continent.

Large, terrestrial animals can't make the same trips. They can't power their own way there like flying birds or bats, and they're not suited to floating along like a screwpine seed. Islands that don't already have large carnivores on them are usually devoid of such voracious creatures, allowing species that could not exist elsewhere—particularly flightless birds—to evolve and thrive. South America is too big to be a haven free of flesh-eaters, proficient hunters will evolve among both mammals and birds, but the continent will nevertheless see life unfold along a unique path, one not only influenced by the creatures carried along since its split from Africa but those that fly or float to the austral expanse. It's not truly isolated, but the available routes are few—a set of filters that will turn South America into a land of terror birds, giant armadillos, and monkeys unlike those anywhere else on the planet.

Moment by moment, the wobbly raft is getting closer to

the South American shore. The day's first light hasn't broken yet, but is promised by a low glow against the eastern horizon that the monkeys have sailed over day after day. To the west, a blurry shadow of something low and dark is beginning to come into view. By time the trio arrives, the sun should be warm enough to dry out their fur and hasten their escape into the forest growing along the coast, a place full of unfamiliar plants and creatures. The tiny, insect-busting teeth of the monkeys will serve them well, though, working just as well on a South American beetle as an African one. It's not the home they know, but they can make it one.

10

Seas of Grass

34 million years ago
Nebraska

WITH EACH PASSING DAY, THE DAWN OF THE CENOZOIC IS coming to a close. The humid, dew-dappled forests that have cradled mammals, birds, reptiles, and so many other Cretaceous survivors are shifting now, the greenhouse world becoming one where a persistent and sweltering summer is feeling the aches of seasonal change once more. As the climate changes, so do the plants and all that rely upon them. Previous diversity is no guarantee of success as low-growing grasses seem to push against the boundaries of dense forests, portents of a change that will reshape the history of Earth's beasts.

Megacerops stands as tall as a short tree. Or, at least, a tall shrub. Eight feet at the shoulder, the bulky and fuzzy herbivore stands among its dinner along the edge of a gently sloshing stream. The creature is like a rhino, but not quite. Like rhinos, *Megacerops* has three toes on each foot with the majority of its body weight balanced along the middle digit. And the mammal's long, long head with comparatively small eyes are also rhino-like. But instead of a pointed horn made of compressed hair, *Megacerops* has a Y-shaped horn jutting up from its nose like a huge, bony slingshot. At this point in time, it's one of the largest land mammals around. No wonder *Megacerops* and its relatives will one day be named brontotheres, the thunder beasts.

Megacerops and their relatives evolved as woodland creatures, munching on soft leaves and fruits with wrinkled, low-crowned molars. The herbivores behave almost like tapirs, their broad feet splaying to support their mass on muddy ground as they browse. At this point in time, however, the forests that have supported the evolution of the brontotheres are turning into something else. The patch of forest the *Megacerops* is enclosed within is still big, stretching for acres from the edges of a riverbank across the lowlands, but a bird's-eye view would reveal it as more of a large island that is slowly closing and constricting.

For tens of millions of years, from the time of *Titanoboa* until now, a warm Earth has hosted dense and broad forests full of flowering plants with sumptuous fruit and thick, delicious leaves. The skeletal anatomy and teeth of the mammals living in these woodlands have reflected their local and lush menus, their chompers low-crowned and their feet best suited to supporting themselves on mucky ground. Within these habitats, primates leaped from branch to branch, bats fluttered after insects in the air, and herbivorous mammals of every shape and size evolved among the seemingly never-ending salad bar. Now the lush heyday is beginning to close, nudging the beasts of these forests to adapt, move, or go extinct as so many other species have before them.

Unaware of these broader changes, *Megacerops* continues to chew on a hackberry tree growing on the edge of the forest. The tree's trunk grows straight up into a broad splay of branches full of almost heart-shaped leaves. The tree doesn't put out its fruit in the height of summer, but late into the fall when dark purple berries appear among the branches. It's too early in the season now for such fruits to appear, but that doesn't dissuade the large herbivore from sampling the foliage.

Picked apart, a mouthful at a time. The hackberry's felt this before. Leaves and stems all torn away with a pull and a slice that shakes the branch being torn apart. It's happening fast, clearly different from the harsh winds of a thunderstorm or even the persistent chewing of caterpillars.

Plants can be surprisingly sensitive organisms. They've had to be. For a tree like the hackberry, rooted in one place for the entire duration of its life, there's no way to physically evade hungry herbivores or forcefully shoo them away. In such circumstances, plants have had to evolve ways to both detect danger and, if not drive the attackers off, quickly repair the damage done from so many hungry mouths. Sensitivity to vibrations is one such warning system, not unlike the way other trees can pick up chemical signals that there are herbivores around.

The hackberry has no sense of what a *Megacerops* is, nor how to defend against such an attacker that's taking entire leaves with each mouthful. Most of the plant's antagonists are tiny: caterpillars and chewing insects that snip away at each leaf and tiny, cicada-like psyllids that leave their eggs on the tree. Those bothersome animals, the tree can usually handle. When attacked, the hackberry can send hormonal signals to cease growth or wrap the pesky psyllid eggs up in a gall to limit the damage. But this? The vibrations of each tooth puncturing leaves and the mammal's pulling at the branches tell the tree that what is happening isn't some storm or other relatively benign event. The tree can distinguish between when it's being denuded of its leaves and when a breeze is just a little rough. Time to give *Megacerops* the hint the mammal isn't welcome.

Like many plants, the hackberry can increase the number of phytotoxins in its leaves in response to herbivores. All those delicate vessels carrying waters, sugars, and other resources for the plant can also be utilized as a defense system, a communication network that signals parts of the plant to start increasing phytotoxins that will—hopefully—be unpleasant for *Megacerops*.

This time, the defense works. Phytotoxins produced by the plant and concentrated in the leaves start to make each mouthful seem a little irritating to the mammal, like an unpleasant spice that irritates the herbivore's throat and nose. It snorts and, finishing a last mouthful, moves away in search of a palate cleanser among the other plants growing along the forest's edge.

Naturally, not every *Megacerops* is equally sensitive to plant toxins. Sometimes a tree can signal *go away* as forcefully as possible and yet the herbivore will keep eating, unbothered. *Megacerops* that are less sensitive to the plant toxins may be more efficient at foraging, leaving more energy to court and mate, passing down the variation to their offspring. Plants that produce greater quantities of toxins, or have other defenses such as thorns, might fare better, beginning the cycle all over again. The back-and-forth is not so much an arms race as a dance, with each partner adjusting to the other's moves as new variations arise and are tested through these everyday interactions.

Plants have been evolving defenses against their predators for a very long time. Stinging hairs on stems, astringent tannins in the leaves, and stashed resources to grow back faster after catastrophe are among the many different, overlapping responses that herbivores of all sizes have prompted plants to evolve. The daily damage and even worldwide catastrophes that plants have survived have brought about their incredible resilience. Arthropods have always been the greatest threat to plants, simply because enough tiny appetites can nevertheless strip an entire tree, but now big herbivores can once again do in minutes what it would take caterpillars days or weeks to accomplish.

Earth hasn't seen an herbivore this big in a very long time. More than 32 million years earlier, at the close of Earth's great reptilian summer, the average animal wandering the conifer groves and floodplains was about the size of this horned beast. A surprise visit from an asteroid permanently changed that. When mammals poked their noses out of their cool burrows and other hiding places into a charred and smoldering world, the largest beasts were only marginally bigger than a house-cat. All those years thriving under the feet of the dinosaurs had allowed mammals to keep a clawhold as all the feathery and scaly giants perished along with fully three-quarters of multicellular life. Luck made the difference between the living and the dead, starting from scratch in a world that had not been so deeply wounded for 135 million years.

Forests grew thick with the giants gone. There were no megaherbivores to push over trees or trample down broad, sun-dappled paths through the trees. The humid swaths of woodland became high-density housing for mammals, an untold bounty of vegetation that acted as the crucible for ongoing mammal evolution. Generalists that survived on whatever they could began to specialize, changing in shape, size, and attribute to break open new niches in this fertile ground. Moment by moment and life by life, mammals began to expand beyond the tiny sizes that still skittered around the forest and racked up new adaptations. Hoofs, in particular, were all the rage, with the ancestors of horses, rhinos, and tapirs faring especially well.

Perissodactyls aren't hard to spot at a glance. From their earliest days these mammals have maintained the tradition of standing on an odd number of toes, but often opting for three rather than the ancestral five. As early camels and the distant ancestors of antelope settled into a life on two toes per foot—marking them as artiodactyls—perissodactyls kept their center of balance along a thick central toe tipped in a tough hoof. The smaller toes to the side helped the mammals keep from slip-sliding around on the mucky ground common to these wet and humid times, splaying and supporting them as they pushed their way through the undergrowth. The early

Cenozoic world seemed to favor their way of doing things. From extremely similar ancestors, tiny creatures that nipped at succulent leaves and sloppily snacked on low-hanging fruit began to split into an array of beasts that would follow the rhythms of Earth's ever-changing greenery. In addition to early horses, tapirs, and rhinos, there were chalicotheres—clawed foragers that looked like horses doing a giant sloth impression—and brontotheres, the thunder beasts that began flourishing about 16 million years after the asteroid impact.

The first brontotheres were average by mammal standards. Trotting through the forest, the largest were about forty pounds. Eocene forests were brimming with similar beasts, all competing with each other for leaves and fruit accessible close to the ground. Only the tree-climbers, including chisel-toothed primates that chattered and leaped through close canopy cover, could enjoy all the vegetation growing more than a couple feet above the ground. No surprise, then, that brontotheres began to grow up. Making the most of those cushioned feet, the little nibblers began spinning off bigger species. While the process required some adjustments—for mammals, getting bigger requires longer gestation times between each generation—the advantages began to kick in immediately. Increasingly larger brontotheres looked less appetizing to the carnivores of their time, dog-like hyaenodonts with long jaws of triangular teeth, and the saber-toothed nimravids, or not-quite-cats whose dark-spotted coats blended in all too well with the thick undergrowth. Few of the flesh-eaters got bigger than a medium-sized dog, often limiting them to prey no bigger than themselves. To a one-ton brontothere, a cat-like predator like the dagger-toothed *Hoplophoneus* would seem about as intimidating as a housecat threatening to scratch. Along with the physiological advantages that come with herbivory—being able to forage on a broader variety of low-quality foods by feeding in bulk, and being able to travel between patches of tasty green more efficiently—there were soon more big brontotheres than small ones. More than fifty different species spread around the Northern Hemisphere, and over half of those were giants weighing a ton or more.

Now most of those giants are gone. The carnivores aren't to blame. On the opposite bank from the drama playing out between *Megacerops* and the hackberry, a wolf-like *Hyaenodon* pads along with its tongue jangling from the side of its mouth. The carnivore stops to watch the brontothere a moment, standing as stock-still as a diorama. In other places and times, its relatives almost get to brontothere size themselves— apex predators heavier than any tiger will ever be. Here, though, the would-be menace is about the size of a golden retriever and wouldn't be able to do much more than nip at the heels of an adult *Megacerops*. Such a predator could still pose a dire threat to a young brontothere, a baby starting life about the same size as its ancestors of 10 million years earlier, but evolution had nudged the mothers to cast their maternal shadows over their offspring for long enough that most babies would be too big to bother by time they struck off on their own. It didn't take all that long. All those leaves, branches, and fruit became milk, fatty and rich as it flowed into the mouths of needy baby brontotheres. By the end of the first year, mothers could give their sore teats a rest and use their U-shaped horns to push their little ones off to find their own patches of the forest to munch on.

Weighing more than three tons, there is little on the landscape that actively concerns this adult female brontothere. Only other *Megacerops* pose dire concern. She doesn't compete or joust like the young bulls do, slamming into each other and shoving in a great spectacle of bellows, broken bones, and loose bowels, but sometimes aggressive males lose sight of who their adversaries are. Last mating season, months before, she had wandered too close to the edge of a jousting ground where a too-aggressive male created a viciously disappointing cycle. He treated any passing females the same as the males that had previously pushed him to the margins, charging and butting with equal violence. She'd been browsing on some juicy leaves along the edge of a clearing when he seemed to materialize out of nowhere at full gallop, slamming into her right flank head-first. The muscles and fat of her expansive sides absorbed most of the impact, but one of her ribs faltered and split in two. For weeks after, every breath burned as the two bones struggled to knit the strap of bone back into one piece. In the end, the tissue created a sort of false joint of frothy bone that made her skin subtly bulge over the spot. Ever since, she's perked her ears to the deep-throated calls and thunder-like cracks of tussling males and found places to browse elsewhere.

Those lush places are becoming islands among expanding grasslands. The conditions that have been so kind to the brontotheres have been slipping away season by season. Rainfall might be more or less in any given year, but the pattern— tallied through changes over millions of years—is toward a drier world that has made water-loving plants retreat to the banks of rivers and lakes. Little by little, patches of grass peek up and spread through the forest as the soil itself loses some of the saturation that allowed forests to knot thickly together.

Megacerops flicks an ear toward the squelch of a foot-fall somewhere deeper in the grove she's browsing on. She's not alarmed. There's no carnivoran stink to suggest she'll be nipped at and no deep rumble of another brontothere, but she's simply curious. The noise doesn't repeat, but a small shadow daintily steps through the forest depths just within her range of sight. As she chews, the edges of her thick lips stained green with chlorophyll, a small horse tiptoes through the glade—*Mesohippus*.

The two are evolutionary cousins. While the ancestors of *Megacerops* spun off a whole range of small and large herbivores, early horses have been molded in a different way. *Mesohippus* is comparable to a medium-sized dog, not all that much larger than its ancestors that stepped carefully through Eocene forests where they seemed snack-sized to the carnivores of the day. Somehow, however, they were not all gobbled up and began to settle in along the margins. The dense tangle of leaves and branches in the forests were comfortable enough, but out in the open these horses could *run*. The harder ground that came with the drier climate seemed to work to their advantage, the horses losing that supportive spread of multiple toes on the ground and instead standing on those singular middle toes.

Little *Mesohippus* pauses, one of his front legs held up off the ground. His little side toes flex and relax a moment, the horse's little ears tilting just to be sure. There is still a whiff of *Hyaenodon* in the air, and he doesn't have the size advantage of *Megacerops*. Standing still, sniffing and listening, he is barely two feet tall at the shoulder. Even at full size, he's smaller than the apex carnivores whose striped coats help them hide among the shadows of the understory and he lacks any horns, antlers, claws, or even a thick hide to deter those drool-coated jaws. But, for the moment, all remains still and he daintily steps a little closer to the juicy forage that its supersized cousin has been enjoying.

The fact that the horse is practically propping itself up on one toe on each foot is a sign of how the world is changing. The very first mammals of so very long ago, snuffly little insectivores that scrabbled after bugs as the very first dinosaurs were getting their feet under them, had five fingers and five toes. This is the ancestral state for all mammals. Some lineages kept that count, but others, naturally, have modified the basic formula. For horses, the story has been all about reduction and changes that only could have manifested at small size.

During the early part of the Eocene, about 10 million years before this moment, the ancestors of *Mesohippus* and *Megacerops* were almost indistinguishable from each other except for the bumps and ridges of their teeth. And they had close to the ancestral number of toes on each foot. One of the first horses, rightly named *Eohippus*, had five toes on each front foot and four on each hind foot. Those little padded feet were great for moving through wet, soppy environments where toes that could spread were great for maintaining a grip in the muck and the mud. But those humid, lush days have been getting ever drier. Slip-resistant feet haven't been as important, especially as carnivores like the hyaenodonts have evolved to ambush and slice up tiny herbivores. Some of the brontotheres circumvented the problem by getting bigger, still spreading their weight over a count of three toes just like the tapirs, rhinos, and other relatives of their time. Horses stayed small, but they began to move faster.

Drier, packed soil behaves differently than mud. Firmer ground opened an opportunity for horses to do something different. They reduced their number of toes to three on each foot, just like the brontotheres, but in time they will go even further. Even on *Mesohippus*, his two side toes— technically the second and fourth toes out of the original five—are much smaller than the central one. They're not nearly as important for bearing the small horse's weight or spreading his weight out. Bearing the weight on a single, central toe might not be as stable as the previous arrangement, but it certainly allows for more speed—especially with some altered limb proportions.

Early horses stood on multiple toes. Over time, the spread of grasslands nudged the ancient equines to single-toed runners.

In broad strokes, how fast a vertebrate can run has a lot to do with the relationship between the upper and lower leg. It's a basic mechanical puzzle that evolution has been working around for hundreds of millions of years at this point. Animals that have very long leg bones above the knee but very short leg bones below the knee move more slowly, partly because their upper leg bones move in massive swings but there is very little of the lower leg to contribute to the animal's reach with each step. It takes more time and muscle power—even if we're just talking about an extra second, or a little more oomph—to move legs like that back and forth, making the upper part of the leg do most of the work. *Mesohippus* isn't like that. His upper leg bones, the humeri and femora, are relatively shorter than the leg and toe bones below the elbow and knee joints. He even gets a boost from standing on tiptoe at all times, the equivalent of the bones that make up our hands and the base of our feet extended into part of the leg. Now those shorter upper leg bones can move forward and be drawn back faster, a quicker step cycle, and those elongated lower leg bones allow for greater reach. He's speedy, at least compared to his ancestors. And with harder-packed ground spreading between the forests, the soil can support the pressure of his body weight pressed onto those central toes when he needs to run. He can easily find himself imperiled in a swamp, especially if he were to step into mires where the weight of his body is focused onto those four points, but the world is shifting to allow new ways of moving around—a trend that will continue not only in horses, but also among distantly related hoofed mammals that will arrive at a similar anatomical solution for running fast.

Evolution doesn't set the beat so much as follow it, however. Survival doesn't just depend on what a species is today, but what it might become tomorrow. Making it in the long term is not based upon perfection to the present moment but adaptability to whatever might come next, and sometimes species meet their end for no other reason than the fact that they have evolved into a kind of anatomical or ecological corner from which there is no route out. In this present moment, meek little *Mesohippus* stands in the shadow of the burly *Megacerops*, but being big isn't all it's cracked up to be.

Earth's climate is changing faster than many species are prepared for. Alterations in temperature and rainfall affect the plants, which in turn affects entire food webs. Such shifts are too slow to be directly perceptible but often too fast for many species to adequately respond to, especially large creatures. Mammals, after all, are not like dinosaurs. Part of the reason the multi-ton wonder of *Megacerops* is possible is because its ancestors evolved to retain their offspring inside for a longer span of time. It's part of the cost of placental mammals gestating their young inside their bodies, bigger babies requiring elaborated time to grow. The reproductive trade-off also slows down generation time, and often requires a constrained number of offspring down to one or two at a time, usually requiring at least one parent to stick around and nurture that baby through the critical first year when it seems everything in the world is out to get them. In a shifting world, fewer babies at longer intervals means less variation for natural selection to work upon. A smaller percentage of offspring will have the variations that suit them better to the world than their parents, perhaps too few to even allow the species' survival. The smaller, faster-producing species face the same challenges, but they spin off enough offspring that there's a greater chance of critical variations—a smaller body, a longer leg, a longer tooth—that can provide just enough of an advantage to have a greater chance of being passed down and elaborated upon through time. For every single generation of *Megacerops* babies there are several of *Mesohippus*, the little horses changing faster than the brontotheres.

Megacerops is holding on to the forests. Their teeth are still low-crowned and best suited to soft, succulent plants. And when the adults are so large that no predator poses a true threat, there's not much reason to leave their favored habitats behind. It's *Mesohippus*, moving ever farther from the forest's edge, that represents the shape of mammals to come, grazers that will follow the lead of the grasslands. The little horse browses the leaves near their bulkier, horned cousin now, nibbling low down with twitching, whiskery lips while *Megacerops* takes mouthful after mouthful of leaves. They share so much, yet are already so different from each other. For now, *Mesohippus* stands along the edge of the larger mammal's shadow. Soon, however, the descendants of this *Mesohippus* will stand in the sun among a sea of grass full of new demands and new possibilities.

11

Partners in Pollination

17 million years ago
New Zealand

THE BAT HAS NO FINESSE. FLAPPING, FLUTTERING, FALLING,
the airborne mammal lands on the dangling flower with more
of a flop than a swoop, flailing with hooked thumb claws to
keep hold. She wants what's inside the flower. It's simply too
good to miss, and who knows how many more nights the
sweets will be on offer? Holding on with those sharp little
thumb claws, she clumsily climbs toward the fuzzy flowers
drooping from the side of the tree to nuzzle her face within
the petals.

The fuzzy, flapping mammal is *Mystacina*, a rainforest bat who is now well-coated in pollen, the sticky little grains hooking into the soft fur of her back, belly, and neck. The grains have evolved for exactly this purpose. Falling to the ground here wouldn't do much good for the flower that's created so much yellow dust. The pollen has to stick. Each one of the microscopic specks is coated in spikes meant to ensnare themselves as best they can on each animal that comes to visit. They're especially effective on mammalian fur, tucked into the mats of filaments covering the bat's warm little body. Each one is a chance for this flower to connect with another and spur new growth among the semitropical forest. But it's not the pollen the nighttime flyer is after. Like most pollinators, the bat has no real use for the tiny grains themselves. "Pollinator" is only the description of an accidental job. Some bees and beetles eat pollen, but bats do not. The fact she's covering herself in the sex cells of another species is just the cost of sipping on a sweet and ephemeral drink.

In about 17 million years, the plant she's hanging from will be known as a kahakaha. The plant doesn't root into the ground. It twines with other plants, anchoring itself into the tissue of living trees to drink up the resources its host creates. That makes the kahakaha an epiphyte—literally a plant growing upon another. It taps into the tissues of its host plant just as that tree sends its roots threading through the soil, sipping from the largesse of the bigger plant. The plant still relies on photosynthesis, but now one step removed. And the host tree can't do everything for the epiphyte. Reproduction is the kahakaha's own affair. The plant has to attract pollinators so that the kahakaha's pollen can be spread among others and kickstart the next generation, and so the plant has evolved to advertise. Dangling from among the thin, wispy leaves of the plant are densely packed tubes of flowers. What the bat wants is within. She wants the nectar.

Flying is a high-energy activity. Bats, in particular, don't soar like birds. The fortunate dinosaur descendants have bodies full of air sacs and hollow bones that reduce the requirements for getting aloft, making them lighter without sacrificing strength. Bats have no such advantages, and the fact that these mammals evolved to fly is even more of an impressive feat than that of the feathery dinosaurs. The limits on bats are much more stringent and have been ever since the mammals evolved the ability to fly during the Eocene, more than 40 million years before this present scene. Bats have to constantly flap their wings in order to stay in the air and reach the treats, an activity that's so draining that it's no wonder little *Mystacina* spend much of their time hobbling around on the ground after insects found on the forest floor. On an island with no large predators, there's no reason not to crawl after food when a crunchy bug skitters by.

But nectar is well worth the effort. A fluid ounce of nectar contains about seventeen calories, just shy of the approximately twenty calories a bat needs every day. And for the earliest parts of bat history, the mammals met those caloric needs by chasing insects through the air. The first and only mammals to truly fly, bats evolved during the early Eocene as part of a great bestial evolutionary burst. The airborne fuzzballs couldn't echolocate, not at first, but rather chased down their prey by sight. The very same genes that helped bats develop their wing membranes *in utero*, however, also caused asymmetries in the skull that helped bats detect where their

reflected squeaks were coming from. Chirps and clicks became like acoustic whiskers, letting bats know where the
crunchy little meals were and how to best intercept the fatty,
gushy meals. Their newfound ability allowed bat evolution
to truly take off, and the mammals have no qualms about
supplementing their usual diet of insects with all-too-brief
flowers full of sweet liquid. Flowers, unlike insects, aren't
prone to running away or squirming under a rock. The
limited-time offer of the flowers is too good to miss. Not
to mention that kahakaha and other flowers evolved to be
more enticing than a crunchy katydid. Nectar is sweet, its
sugar pleasant to chiropteran tastebuds as well as calorically
rewarding. Plants need to make relatively little of the stuff
to get the reward of the bats doing their reproductive dirty
work.

The sought-after fluid in each little kahakaha flower is
kept in a nectary, deep inside. Each drop secreted by the
flowers is full of sugar, a reward for attentive pollinators.
After all, the bat can't just drag herself over to slurp from the
petals and take off. The nectar is part of a trade that the bat
doesn't know she's making. Fast as this little evolutionary
dance plays, the bats have to stick their faces and tongues
deep into each flower, a distraction while the mammal inadvertently covers herself with pollen. On especially thirsty
nights, she is positively dusted with the stuff—a sticky yellow instead of the usual mammalian brown. She ends up
looking like a neon-colored corn puff with wings, the botanical dusting taking quite a bit of grooming to remove.

The bat and her roostmates in this part of the dense, humid forest are far from the earliest pollinators. They are simply

among the latest creatures to take up this special relationship with plants, one that preceded the evolution of angiosperms by tens of millions of years. The partners change but the tune remains the same.

Pollen is incredibly ancient. More than 260 million years before the bat's nighttime visit to the flower, back in the days when sail-backed protomammals were crawling over the planet, earwig-like insects munched on pollen-packed buds of ancient gymnosperms with thick, strap-like leaves called cordaitales. The invertebrates were looking for meals, nibbling into the plants' store of sex cells, but of course they were messy eaters and ended up carrying pollen between their stops for snacks. Such interactions didn't begin a specialized and intimate relationship between the plants and pollinators, however. The pollen-eating insects were relatively few and specialized on particular plants. The early cycads, ginkgoes, and other gymnosperms were better off waiting for opportune wind gusts to blow their pollen into the air in the small chance that the grains would land on the appropriate part of another plant. Animals could offer some supplementary assistance, but it was nothing to be relied upon. Gymnosperms, for the most part, seemed incapable of evolving the kinds of flowers or enticing lures that would interest insects, birds, mammals, and other potential pollinators to visit them. It really was a matter of chance. At least until angiosperms began to evolve colorful, fragrant, nectar-oozing flowers. With these innovations among the evolutionary newcomers, the relationship wasn't just one of two organisms happening to bump into each other. Plants could bring their pollinators to them.

Bats have been among Earth's most important pollinators
for millions of years.

Developing such close relationships always comes with risks. Some long-term survivors have avoided developing such reliance on other species. *Ginkgo*—a single surviving species of what was once an incredibly diverse family, a group that underwrote the bulk of the biggest dinosaurs—is pollinated by the wind. Insects wandering over a plant that bears pollen cones might blunder onto another ginkgo that has ovules, but such occurrences are unreliable. There are going to be more friendly breezes than helpful insects. Remaining unspecialized allowed *Ginkgo* to survive the world's worst catastrophe and persist for millions and millions of years more. But many angiosperms have opted for something different, interactions with their pollinators that has fueled a burst of colors and scents that require a vibrant, precarious existence.

A flower that's easily spotted and relatively open in the center can attract a wide variety of pollinators. Beetles, bees, birds, and more may visit, picking up pollen and spreading it around. The offer isn't without risk to the plants. Nectar requires energy and resources to produce, and pollinators best able to lap up the sweet fluid have a better chance of surviving and passing down their minor specializations to their offspring. Flowers might get a great deal of attention for little in return. Some animals, too, may simply eat the flowers, undermining the entire process. Some flowers, then, evolve to be more secluded with harder-to-access nectaries that require animals with only just the right bodies or mouthparts. It's a specialization in which plant and pollinator evolve together, becoming reliant upon each other. The paired species alter each other through the years and millennia, their very features evolving to match each other. Petals and nectaries rearrange themselves to only allow a certain stretch of tongue, a game of evolutionary tension in which the most precious parts of the flower are held almost impossibly out of reach but not quite, a gift that can only be given through an evolutionary compromise that demands the pollinator devote themselves to this specific plant. The species become more than acquaintances. Through countless impulses and happenstances made before they were even born, they are made for each other. And, for a time, such adaptive bonding can seem like a safe bet. A flower might have a single pollinating species out there, one with mouthparts specifically evolved to reach the nectar within and leave with a dusting of pollen. Such an evolutionary pact is not a promise of support, or even survival. Pollinators play other roles in their ecosystems, and sometimes they become extinct.

In the same forest where the kahakaha grows, lianas wind their way up tree trunks and drape across branches as they grow over the larger plants to reach the daylight. One of them, a relative of the screwpines with their floating seeds, had a close relationship with one of the bat species in this tropical forest. The woody plant puts out triangular flowers with finger-like, pollen-coated projections inside—perfect for flying creatures to come stick their faces in and get dusted. And yet the bat who most frequently visited the liana has become extinct, leaving the flowers almost wholly unattended for the occasional bird and insect.

If the liana had become more specialized, evolving structures to only allow the bat's muzzle close to its nectar, it would be painted into an evolutionary corner. Unless another animal had similar specializations, the plant, too, might disappear with its pollinator. Luckily the liana was merely favored by the long-lost bat, available to be pollinated by other creatures if only they might stop by. There are enough birds in these forests that even their occasional drop-ins are sufficient to help keep the plants pollinating each other and hanging on, but a more enthusiastic pollinator is required. Another bat species, like *Mystacina*, would help a great deal, or perhaps a small climbing mammal like a possum. Only time will tell. A friendly breeze can only be of so much help to the stunningly orange flowers, waiting for their sweet invitation to be requited.

12

'Nip Trip

9 million years ago
Western North America

HER TEETH WERE NEVER MEANT FOR PLANTS. SHE STILL sliced through the leaves and stalks, anyway, her sandpaper-rough tongue gently flicking as the mashed green matter sticks to her gums, the scissor-like action of her cheek teeth slicing through the cell walls. The back edge of her serrated saber started to pick up some of the shredded pieces, smeared with green. Even sabercats need a treat sometimes.

One of the largest saber-toothed cats to ever prowl, she's a *Machairodus*. Her seven hundred pounds of flesh and bone have been built and fueled by the tissues of other creatures, the herbivores that have converted the abundant plant matter of the surrounding grasslands and woodlands into digestible meat. Large prey is her specialty. Her anatomy requires it. Her broad paws support burly arms, of course including retractile claws more than capable of hooking into the soft flesh of the unwary horses and camels she has pounced upon. And her long, curved fangs have evolved to cause incredible trauma in an instant—ripping out throats and slicing bellies, opening bodies so that blood loss and shock might make the final bite easier. The skull of her species has been adapted to support her peculiar method of killing and carving, two flat and elongate canines jutting down from her upper jaw and generally contributing to her persistent resting cat face. A thick, whiskery muzzle partially obscures them, leaving an inch or two of fang visible just below.

The cat can even crunch bones when she needs to. In drier years, when herbivores scattered in search of suitable water holes and greenery, the plant-munchers became fewer and farther between. She had to make the most of each kill that she made herself, happened upon, or stole from smaller competitors. When the muscle and viscera were gone, she'd hold down the thick, long bones between her grasping paws and gnaw at them until the hardened calcium gave way to luscious marrow inside. In better years, she would often start with the juicy muscles of the legs, rump, and shoulders before picking through the organs and scraping off whatever other morsels she preferred. She'd stay on a meal for as long as she wished, sometimes reducing it to scattered and gnawed bones and other times leaving other scavengers to pick over the lesser cuts. Impressive for a cat who's not even fully grown yet. If she survives to the end of her growth spurt, she'll be about nine hundred pounds of cat. And because of that, she's hungry a great deal of the time. When she's not dozing in the shade, she's usually hunting. Find a meal, consume what she can, and move on, snoozing through the heat of the day until the concealing shadows begin to grow long again.

Plants don't make up a regular part of her diet. When she ingests them, it's often as debris or partially digested glop from the bellies of her kills. Still, there is one bit of greenery she's developed a particular fondness for. Catnip, when she can find it, is always a welcome treat.

Cats are not exactly pollinators. Nor are they voracious plant predators. The relationship between the tiger-sized, fanged cat and the mint-like herb is not nearly as straightforward as that. The relationship between cat and catnip is an accident, two separate trajectories that somehow, out of all possibilities, came to intersect in a wonderfully pleasant way—at least for the felid.

At a glance, catnip is just another unassuming part of the forest undergrowth growing low down near streams and other places with adequate water among these Miocene woodlands. Its oval, serrated leaves and purple flowers give it away, but the plant doesn't bear any conspicuous berries, grow to impressive size, or much else that a passing sabercat would find remarkable. Such low-growing plants usually serve as little more than scent billboards when the cats spray them with urine to mark their presence. And given that sabercats are hypercarnivores— focused on flesh more than anything else—there's not much reason for the cats to seek out the greenery just for its nutritional value. Cats prefer plants that have been converted into muscle and guts. And despite all this there is still something in the catnip's scent, a compound that causes cats to flop, roll, rub their cheeks, and be anything other than fearsome. That compound is nepetalactone.

Insects have shaped a great deal of plant evolution, part of an ever-busy, hidden world beneath the paws and hooves of the big Cenozoic mammals. Even now, in this otherwise peaceful vista, the relationship between plants and insects is held at a tension point. Many insects are hungry for plants—their slicing, crushing, drilling, pulverizing mouthparts capable of bringing down entire forests if left unchecked—and plants are constantly evolving new ways to dissuade insects from munching on them. No individual plant needs to be impossible to consume. They just have to be too much trouble—a little too tough, a little too toxic, a little too challenging to bother eating when there's easier fare just over *there*. Nepetalactone is the result of this eon-old back-and-forth, a biological compound that became catnip's way of saying "Why don't you go bother someone else?"

When a plant creates compounds in response to animal predators, however, those compounds often reach unintended recipients. Such unforeseen effects have evolved plenty of times before. The venom of snakes evolved to help the reptiles kill prey quickly and more efficiently, for example, but that doesn't mean that a bite in defense won't make a curious bear dog terribly sick. Catnip is much the same. The active compound fits into receptors in the felid brain keyed to an altered behavioral state, an effect the cats enjoy a great deal. To a mosquito, nepetalactone is annoying. But to a cat, the compound is more than simply pleasant.

The reason a plant could make a compound that affects insects as well as giant saber-toothed cats can be drawn back to evolutionary changes that happened hundreds of millions of years before. It all comes down to a set of sensory nerves called TRPA1 that are about as old as animals are. The neuron is a kind of alarm bell for animals, not transmitting pain but a warning about stimuli that are too hot, too cold, too itchy, too irritating. It's part of an alarm system that tells many animals to move away from whatever's causing the unpleasantness. For insects from flies to mosquitos, nepetalactone sets off TRPA1 like a firecracker. The flies won't lay eggs among the catnip plants, nor do mosquitos seek their blood meals nearby. The plant evolved to tell one of its direct predators to stay away, but because the neurons it targets are so widespread it works on all sorts of insects. And while catnip makes many insects bug other plants, the cats enjoy more of a figurative buzz.

Machairodus is just a few chews into the swath of low-growing catnip before she leans into a deep stretch, pushing her chin low to the ground as her massive mitts slide through the clumps of catnip. It just feels so good. Tilting her head to the side, she begins to rub her striped cheek against the plants, intensifying the sensation as well as rubbing her scent all over the plants. With a twist, she rolls and flops her spotted side on the ground in a heavy thud, pushing with her feet and seeming to swim through the bright-green groundcover. She can't help herself. She just has to loll and roll, cheek-rubbing and purring in the dappled shade of this very good spot. Her deep rumbles provide rhythm to the shifting shade of the day.

Cats' fondness for catnip is an unintentional effect of plants
developing compounds to defend themselves from insects.

Machairodus hears a sound from just beyond the tree she's
shading herself under. It's not quite a woof, but it's another
deep, carnivoran sound. A greeting. Another *Machairodus*,
much like her, down to the shock of white at the end of her
stubby tail—a flag that helps the cats understand each other's body language. In other circumstances, like a fresh and
sloppy kill, the approach of another sabercat would almost
certainly result in bared fangs, hisses, and claws-out swats.
But she knows this cat. They've been hunting together for
weeks now, eating a little less off each carcass but feeding
more often thanks to their cooperation. The second female
purrs as she approaches, a deep and resonant rumble making
her nostrils flare a little with each pulse. The first rolls on her
back, splayed paws reaching out to grab the other female,
mouthing and gently rabbit-kicking with her back feet like a
giant kitten. She's been into the catnip for more than a few
minutes now, the effects having fully taken hold.

The second *Machairodus* is just as enthralled by the catnip as the first. She sniffs, awkwardly hanging her mouth open so that the invisible compounds of the plant travel up into the roof of her mouth to a special organ that detects the smelly subtleties of life. This is good, fresh catnip. She flops her feline bulk down, kicking up a small cloud of dust as she rubs her muzzle against the plant, crushing some of the leaves in the process. The other cat, seeing an opportunity, play pounces, growling and rubbing her broad head against her partner's neck and chest, sharing the scent that marks what in this ancient world they consider to be their own. Batting, biting, and rolling, the carnivores play like kittens until catnip's other notorious effect—drowsiness—begins to sink in. The cats fall asleep against each other rump to rump, their noses pointed in opposite directions in case some unforeseen intruder decides to interrupt their pleasant doze beneath the lilting shade.

The sunlight shifts and slides across their spotted pelts, long and wavering stripes of black contrasted against coffee-with-cream brown. If you were to shave them down to their skin, you'd see the same paintbrush blotches of black beneath all the fur. The haphazard pattern is no evolutionary accident. Yes, the width of tear-like stripes on their faces and the precise tangle of horizontal spots along their flanks help the cats recognize each other, but the splotches and blotches have graced their fur because of the plants the cats so frequently stalk among. The change goes deep into the felines' genes, each embryonic kitten inside her mother painted in a unique array of splotches thanks to regulatory genes that determine her pattern.

Cats don't gain their colors through simple inheritance. Two closely related genera or even species of cat can wear strikingly different colors. It's the environment—the places that they stalk—that give cats their coats. Plants set the palette. The cats are simply following what will help them blend in best.

If the dozing *Machairodus* were cats of the open grassland, dragging their bellies against the pebbly soil as they stalked unwary camels and horses, they'd likely wear cloaks of buffed brown. Grass spends much more of the year in its dried-out golden state than green, and the vast fields of it can provide excellent concealment—even for cats as large as *Machairodus*—if you're the right color. Brown, tan, and gold are the colors of cats who live their lives out in the open, bounding over the flats to tackle baby mastodons who wander too far away from the herd.

But *Machairodus* are stealthier hunters. They are among the great "dirk-toothed" cats, like the iconic *Smilodon* that will follow them as the world tips into its Ice Ages. Dirk tooths don't chase their prey, or at least tend to give up after a few yards. They are quiet, powerful hunters with shoulder and arm muscles that tense under their pelts with every padded step. Woodlands suit these cats better, places where there is always another tree to hide behind and the nightly darkness creates a maze of tree trunks. The reflective layers at the back of their eyes—a rind of tissue called the tapetum lucidum—bounces the available light back, even when the moon is just a sliver of a Cheshire Cat grin, granting the sabertooths the ability to see their groggy prey who might not see them. And even in those darkened hours, a dappled coat of elongated spots breaks up their outlines, turning the cats into shadows whose prey might only detect them by their rank odor or the unintentional snap of a twig.

Curled up in a post-catnip haze, the *Machairodus* are

hardly a threat to anything at this precise moment. But when they awake, with bellies that feel annoyingly empty, their coats will help them blend in against a forest floor that seems spotted with shade almost as they are.

13

Far from the Tree

6 million years ago
Ethiopia

IT'S TOO HOT IN THE SUN. EVERYTHING SEEMS SMOTHERING and still in this part of the late afternoon, the day's heat at its peak even as the sun has been slowly rolling toward the west for hours now. All *Ikelohyaena* can do is sit in the shade and pant, radiating heat off of his belly and striped backside as he stretches out and lets his body lightly jostle with each rapid breath. At least there's entertainment. Just there, along the edge of a long branch stretching beyond the border of the forest, are two hairy primates fussing with each other.

On the ground, the apes might make a tasty snack. They only stand about four feet tall and they lack almost any sort of defense against bone-crushing jaws. The apes don't have any armor, claws, or blade-like canine teeth, and they're absolutely awkward when moving on the ground. They can't run, nor are they particularly skilled at walking on all fours on their knuckles. Legs splayed, they do a sort of shuffle, always ready to hobble back toward the nearest tree, where their grasping feet and long, curved fingers are better suited to moving through the open canopy. *Ikelohyaena* keeps watching the two along the long branch of a tree swat and screech at each other over some piece of green food in the hopes that one might tumble. They're not as meaty as an antelope, but at least their soft parts are relatively easy to gnash into.

The pair of quarreling primates are both *Ardipithecus ramidus*, members of the same species. At a brief glance, they're not all that different from the apes of a million or even five million years before. What makes these thick-haired primates special will only be recognizable in retrospect millions of years later. There is nothing in these woodlands that recognizes *Ardipithecus ramidus* as some of the first humans to dwell on the planet. Nor can these hominins perceive that global changes will gradually nudge their descendants out of the trees, the future of their family reliant upon traits that evolved among the leaves and branches.

The forests that have given primates their characteristic features have been changing for some time now. A planet that was once a primate paradise has been requiring some significant changes from the grabby, chattering mammals. In the earliest days of the Eocene, when primates began to truly thrive, the mammals were rapidly evolving and spreading through ancient forests nourished by a warm, wet, and generally more homogenous global climate. There were not marked dry seasons and wet seasons. Vast forests covered large expanses of the planet, primates evolving beneath the canopies as each new species opened a different niche among the leaves. Since that time, however, Earth's climate isn't so even or rain-soaked. Earth has become drier, seasons taking hold in various parts of the planet. Weeks or even months might go by without a single refreshing drop of rain in the warmer regions. Then everything that was needed in those parched months will arrive in a torrent that causes rivers to jump their banks and floods to sweep across the lowlands. Great swaths of forest remain, but they're becoming more dispersed. Between them are slowly growing open spaces carpeted with grasses that are more efficient than other plants at sucking carbon dioxide out of the air for photosynthesis. These are C3 grasses, plants that responded to lower carbon dioxide levels in the atmosphere by evolving a better way to access the compound only to flourish when levels of CO_2 went back up again. And these grasses are better suited to a life of feast and famine, growing quickly in response to the wet and spreading their seeds as quickly as possible. When the dry months start to sear the ground, the grasses have always made their play. Trees and other forest plants live longer, but they are much more dependent upon water sources and nourishing soils.

The world has generally grown cooler since the days of seemingly endless canopies, too. Forests that once extended almost across Africa's midsection, from the Atlantic Coast almost to the Indian Ocean, have been shrinking back and thinning around a ring of grasslands. Ice has been building up at the poles again for some time now, slowly pushing the Earth toward what will be incredible glacial cycles when vast sheets of ice expand and retreat to only play the cycle out again. And as global temperatures have gotten cooler, the climate has also grown drier. Forests that thrived on regular rains have fundamentally changed, what was once evergreen shifting to a new annual palette of green-and-brittle-gold with the back-and-forth of dry and wet seasons. Semitropical forests have been split into patches of woodland with grasses between, meaning that many primates can no longer count on twining, entangled branches to move above the heads of the cats and hyenas below, and there were not so many lusciously fruiting trees to pluck sweet fruit from.

The forests and the climate have changed over the course of millions of years, yet still fast enough to require prehistoric primates to move to the rhythm. Like all life under pressure, primates could stay put in their dwindling habitats, move to find more persistent patches of forest, or adapt to the shifting landscape, still calling the trees home but spending more and more time on open ground where all manner of carnivores would be curious about the relatively novel morsels shambling through the grass. By the time of *Ardipithecus*, the dwindling species of apes across Eurasia and Africa had split their strategies. The relatives and ancestors of what would one day become chimpanzees and gorillas clung tight to the forests closer to Earth's equator while the earliest humans began to awkwardly wade through the seas of golden grass along Africa's eastern Rift Valley.

Where it can still be found, thick forest is an ideal place for many primates. Remaining forest canopies stacked like arboreal layer cakes rose into the air, green all the year round. When trees aren't producing incredible quantities of fruit, there are plenty of thick, nutritious leaves to eat. As both food and shelter, forests have always held primates close. And in an array of ways, apes have found their own ways to move through the trees and over the ground when necessary. Different apes swung off tree limbs, clambered on all fours over branches, and tottered upright through the canopy while grabbing supports, co-opting those strategies to the ground below when making occasional trips down. Uncomfortable as the apes could be when far from their canopy homes, it was life in the trees that had molded their bodies for that most simple but challenging evolutionary outcome—good enough. Whether it's a tiny red algae growing in the sand or an ape shuffling on the ground, an organism's body and behavior only need to be satisfactory to open up new evolutionary possibilities. In the case of these primates, foraging in groups and being just fast enough to get back to the safety of the trees was enough to allow some species to start venturing farther and farther from the canopies as climate change broke up their elevated refuges.

Ardipithecus ramidus is one of the earliest humans yet identified, still bearing skeletal adaptations to life in the trees.

The *Ardipithecus* seem to know how poorly they navigate the grassland. Or perhaps they sense the curious carnivoran eyes watching them. As much as they can, they stay in the trees. Even as the two humans chase each other, shaking branches as they move foot over foot through the tree and send loose leaves lilting down to the ground, each of the *Ardipithecus* pair are confident in their movements. Primates have had forward-facing eyes and grasping hands and feet for tens of millions of years. Their ability to grab hold of branches was also useful for picking food, snatching insects, grooming, and manipulating everything they could safely touch, a wholly different way of interacting with life in the trees than almost any other form of mammal.

While some monkeys and apes have continued to climb and clamber through the trees, other lineages have been more comfortable taking the risk of venturing out on open ground to forage from the abundance of new, grassy greenery. Ancient baboons and their relatives, especially, shifted to live on the ground and graze among the abundant, tough grasses. Group living has helped such primates make the switch. Big cats, eagles, hyenas, bears, crocodiles, and other sharp-toothed creatures still take their share, but living in groups meant more eyes to keep a lookout and sound an alert if a striped shadow was seen slinking through the tall grasses or a pair of emerald-green eyes peeked out above the surface of a pond. Even if the first primate to yell a warning has more of a chance of being picked off, drawing attention to themselves, their relatives might live on. And so primitive baboons, early apes like *Morotopithecus*, and other primates have carefully backed themselves down the trunks of their familiar arboreal homes to plunk their ischial callosities among the grasses and pluck leaves to their heart's content. The descendants of *Ardipithecus* will be doing the same. Moving in such habitats feels unusual, and those apes that are able to stand up a little longer will better navigate the flats, but being able to sit in great big salad bowls was worth the occasional back pain and leg cramp.

In this evolutionary moment, *Ardipithecus* are living something of a split existence. The remaining trees are home and safety, but they can't count on the arbors to fill their bellies anymore. For food, to cross patches of forest, and even out of sheer curiosity, *Ardipithecus* have to go down to the ground sometimes. It's this simple fact of existence, in this part of ancient East Africa, that will nudge humans to one of the most ridiculous anatomical arrangements of all time.

The two arguing *Ardipithecus* eventually settle their debate, the hard feelings over who saw the crunchy beetle first settled among low hoots and some satisfying tugs of fur as the one who enjoyed the burst of insect flavor begins to groom the other. Other such snacks will surely wander along as they move from one branch to the next. Part of what makes them so comfortable here are their arms and shoulders. *Ardipithecus* not only grasp the trees with opposable thumbs, their long arms giving them incredible reach, but their wrist and shoulder joints are extraordinarily flexible for mammals. *Ikelohyaena* watching down below has the wrists and shoulders of a pursuit predator, best suited to a forward-and-back motion that effectively allows them to pull themselves forward as their back legs push, a hinge with limited motion. *Ardipithecus*, by contrast, can rotate their wrists in multiple directions and grab at all sorts of angles. The shoulders of these humans, too, are loose arrangements of bone centered over the back rather than at the flanks. The hominins can reach overhand, grasping above their heads, pulling branches down, tossing rocks underhand, and all sorts of different motions. They are another example of the general primate trend of trading speed for dexterity. All of this is because of millions and millions of years in the trees, habitats full of strange angles where the next best handhold might require some extra flexibility. Instead of running on all fours, *Ardipithecus* evolved to climb and push off with their shorter, bowed legs, navigating complex pathways far different from the relatively flat savanna below.

Ardipithecus don't turn up their noses at meat when they can get it. A sleepy lizard crawling over the branch is an iron-tinged treat. The humans have the teeth to handle it. Square-shaped incisors nip, short canines pierce, and a series of premolars and molars grind and mash, those grasping hands and a tilt of the head allowing the hominins to place their morsel up to just the right teeth for the task. Molars can crush small bones about as well as they smash seeds. Most of the time, though, they eat from among their bark-covered homes. Fruit and leaves fuel their bodies, and, as herbivores have struggled with since the Carboniferous, at least, digesting vegetation isn't easy. *Ardipithecus* have large, round guts, effectively fermenting vats where their bodies attempt to move all that plant matter through as slowly as they possibly can. To compensate, the hips of such apes are becoming more bowl-shaped to adequately cradle their viscera and the extra weight of their meals, essential for properly balancing while moving around.

In the coming centuries and millennia, forests will pull back even more into dispersed woodlands—stands of trees in seas of grass. The success of elephants, giraffes, and other huge herbivores will help maintain these open spaces just as the great dinosaurs did in the Mesozoic, not only feeding from trees and trampling ground but pushing entire trees over to better nibble on leaves and fruit that would otherwise remain out of reach. Early humans like *Ardipithecus* will simply have to get used to moving on the ground if they are to survive, and there's more than one way to do it. They could become knuckle-walkers, moving on all fours and curling their long fingers to use their knuckles almost like paw pads. It's a way to move like a quadruped while still keeping those wonderful, precise fingers. It's also a way of moving that requires particular anatomical specializations to bear more weight on the wrist, and it comes with its own aches and pains. When *Ardipithecus* come down to the ground, it's often easier to stand

and waddle around. If only they had kept their tails. Like all apes, *Ardipithecus* lacks a tail. All that's left of what their ancestors once had is a crumb of mashed-together bones just below the pelvis. If they had retained a tail, as monkeys have, then it might be adapted into a useful counterbalance and not require that *Ardipithecus* try to stand up straight, putting tension on the muscles of the lower back. Such as it is, the humans need to tilt back to find their balance and spread their legs apart as a more stable base, each step technically being a fall that they catch on each step. And yet, for all the discomfort and lack of speed, it works.

Ikelohyaena, a little less excited now that the squabble's calmed down, yawns and stretches, flexing dusty paws under the shade. There probably won't be a hominin dinner today. It's later hyena species that will have better luck, watching from the grass as humans tentatively step into the world of the carnivores. There is both possibility and terror among the spreading grasslands, a new context that will both mold and test all the ways it is to be human.

14

Ancient Autumn

3 million years ago
Willershausen, Germany

THE ENTIRE FOREST SEEMS TO ASK FOR QUIET, A LONG
shhh among the trees. Another great autumnal gust sweeps
through the woods, shaking multicolored leaves in a susurration so all-encompassing it seems like a soft blanket of sound.
Reds, yellows, oranges, and a vanishing hint of green wave
and flutter, a few leaves carried off to tumble and twirl along
the wind. Each one carries the last memory of summer as
they lilt and eventually settle to carpet the forest floor as autumn makes itself comfortable in these ancient groves.

In ways large and small, the forest is settling in for its slumber. The season of growth is long past. Birds aren't belting their best to impress potential mates. Insects are sticking closer to their hiding spaces. For many species, the year's offspring must now fend for themselves among the increasingly chilly nights. There's no courting or nesting. Life seems to slow as the days grow shorter and shorter, some mornings bringing a rim of frost to puddles and ponds in this part of Pliocene Europe. Some creatures have moved farther south, following the fading edge of warmth and the possibility of vegetation that does not slough off for months at a time. But some have lingered, feeling the seasonal turn. The great mastodon *Anancus* dozing to the soothing noise of the rustling leaves has seen this seasonal swing happen year after year, since she was a floppy-trunked youngster never far from the shadow of her mother.

The blaze of color she's watched settle in each autumn takes its cue from the light. Trees have to live according to the rhythms of weather and climate, varying from one year to the next. One good year might be followed by a hard one, or several, trees slowing their growth and leaving a record within the rings that document their story just as geologic strata record the history of entire habitats. But shorter days, longer nights, and the colder temperatures that come with them are practically constants dictated by Earth's cooler and drier climate. The trees of this Eurasian forest have evolved to move with the changes, suppressing their activities so that they might see next spring. Earth isn't basking in an endless, warm summer anymore. Through millions of years, a semitropical planet has turned to one with polar and temperate zones, places where organisms have to be flexible. Survival means swaying with the seasons.

It's hard to thrive in the cold and the dark. Less sunlight each day means a cutback of photosynthesis, that most essential of plant activities. Trees could keep at it through the chillier half of the year, as evergreens do, but those gymnosperms have evolved to withstand extremes. Conifers have arrays of waxy needles that resist freezing and all the damage ice crystals can do. The comparatively soft and broad leaves of most angiosperms can't do the same. When the deep freezes set in and the ground is cold, such thin leaves would be damaged and torn even if the trees were to keep them. Without evolving new leaf shapes, the broad-leafed trees have done something else instead. Many angiosperm trees reclaim the materials they need from their own leaves, effectively going into a slumber when getting by on less may allow them to vibrantly leaf out again as warmer and longer days return.

The trees still spreading leaves above the head of *Anancus* are not intentionally changing their shades. The display is not an active effort to replace the summertime green with autumnal orange. During the spring and summer, leafy greens are created by an abundance of chlorophyll. They take in most every shade except for green, just as it's been for hundreds of millions of years. But leaves also have other compounds within them, hidden by the abundance of chlorophyll. Many trees have carotenoids, pigments that appear as vibrantly yellow and orange when allowed to stand out. These hidden shades get their time to shine thanks to the process trees and shrubs use to shut down their leaves for the season, gradual and imperfect reabsorption causing leaves to change their colors. Some of the sugars created by the busily photosynthesizing leaves get shut off as the leaves are left to their end, as well, showcasing particular pigments—anthocyanins—that blaze bright red. As the long nights settle in and cooler air temperatures arrive, trees start shutting down the veins and vessels to their leaves as each stops creating new chlorophyll. The greens fade away, revealing shades determined by the carotenoids and anthocyanins that were there all along. It's a final splash of color before the tree begins to once again rid itself of what it doesn't presently need. In this forest, the oval leaves of ash trees turn red, while shrubs called boxwoods, growing wherever the treetops allow enough light, shade to orange. Even though the bright hues make the gray hide of *Anancus* look a little on the drab side, the massive elephant now stands out among all the amazing colors because the subdued palette of winter settles in.

Trees have been signaling the seasonal change since at least the time of *Megacerops* and *Mesohippus*, more than 30 million years before this moment. Even during those warmer Eocene days, plants closer to the poles had to cope with longer nights—sometimes even months of encompassing darkness. Even without persistently freezing temperatures, continuing to spread leaves under darkened skies was as hopeless as a dinosaur trying to breathe underwater. Trees that managed to take root into extreme places evolved a different way to handle the darkness, gradually spreading to other parts of the world as temperate zones spread. The harsh habitats of the poles helped bring about adaptations that have spread outward as Earth has changed.

Big mammals are better able to cope with dropping temperatures by virtue of their greater internal volume, but it's starting to become a little too chilly even for a multi-ton elephant. Time to wake up and get moving. Normally *Anancus* sleeps standing up, propped against the trunk of sturdy trees.

In such moments, hers is just another trunk among the columns of brown and gray. But standing more than eight feet at the shoulder and weighing more than six tons, she doesn't have much to worry about from even the most inquisitive sabercat. Lying down and resting those pillar-like legs feels good once in a while. At least until she's hungry. Now it's time to rise, and she awkwardly pushes her shoulder into the leaf-littered ground to kick her back legs out, levering her body so that she can get her front feet beneath her immense body and press, a great proboscidean push-up. She shakes and snorts, flipping her muscular trunk over one of her long, straight tusks as she comes back into herself.

Anancus is a forest elephant. The massive molars in her mouth can handle entire branches as well as leaves and fruits. Her life is one of gradually pruning the forest, leaving bare patches from the ground to the tallest her trunk can reach. Over these past few weeks she's been noshing her way southward until she reaches more comfortable temperatures. A season endlessly chewing on frostbitten bark alone will not be pleasant, and with every snack she's slowly been moving herself farther and farther south toward forests that keep leaves on offer longer. It's one of the perks of being big. She may walk slowly but she can still cover more ground in a day than most mammals do in a week.

She squishes and sometimes squelches along a multicolored carpet of wet leaves as she moves, a vast area rug laid down in sodden earth tones. It's getting cold enough at night now that some of the leaves are given a glittering edge of frost as the sun rises, and as the held-over moisture melts, it slips between the millions of discarded botanical sheaves. The dendrological rugs building up along the game trails and shaded spaces is so much more than litter. It's a miniature world that only exists for a few months out of the year, when leaves that sprouted to catch summer breezes come back to the soil and invertebrates huddle beneath their accumulated shelter.

The bright colors of ash trees in autumn pop as the plant absorbs chlorophyll and other compounds in the leaves become more visible.

The world of the leaf piles might as well be invisible to the *Anancus*. The elephant has no knowledge of the ever-busy, teeming life literally beneath her feet, not just fungi and bacteria but springtails and spiders, millipedes and queen bees, inhabiting a decomposing realm that is nevertheless vital to them. We need to once again think small to perceive it. To a creature of our size, a harvestman—or daddy longlegs, if you like—is a strange, spindly little animal with a small body and legs that look as thin as thread. But imagine if you were small enough that their six eyes were about the same size as our two, if you were miniaturized such that all the minute details of their segmented body were as clear to us as the exterior of a car. At such a size, leaves are suddenly banners crisscrossed by veins, worms are thicker than fire hoses, and a beast such as the ever-wandering *Anancus* is a moving, shadowy titan that we could only really perceive as trunk, belly, and legs. The topography of the world is suddenly very different, a world of rotting leaves as big as bedsheets and pill bugs that would seem to roll up as big as boulders.

The slick and sometimes brittle world we now find ourselves in is a parting gift from the deciduous trees. In eons past, a warmer world precluded the need for such shelters. Instead of four seasons, much of the Earth's surface experienced two—the dry season and the wet season. Finding water and shade was much more important than escaping the seasonal chill. As Earth's climate has become more variable and seasons have taken on new characters, however, life on Earth has had to adapt to survive an array of conditions from chilly spring rains to scorching summer evenings and iced-over winters. Big organisms like *Anancus* are able to move with the changes, capable of walking miles each day for the resources and conditions their bodies require. Ants and snails cannot do the same. At a speed up to a foot an hour, even a snail with incredible powers of endurance would only be able to pass a tree or two in a day. Those invertebrates that don't perish for the season, leaving behind their eggs for the spring, have no choice but to shelter in place, and the evolution of trees and shrubs that shed their leaves have been among the happiest accidents in the long history of forests.

By itself, a leaf seems to be a fragile thing. Even when leaves are fresh and impossibly green, they're still easy to pull from their perch on a tree. Leaves are easily torn, and, of course, they can do little against the constant onslaught of hungry insects except pump noxious compounds into themselves to encourage their dinner guests to go visit some other plant. Leaves dry out, burn, curl from the onslaught of parasitic fungi, and are sipped to death by insects with piercing mouthparts, subject to almost every form of damage and indignity nature has to offer. You might expect a fallen leaf to disappear as quickly as a sheet of tissue paper in a rainstorm. And yet fallen leaves can remain on the forest floor for a year or more, turning brown and gradually decaying but still persisting much longer than they ever were clinging to their parent tree. Their secret is the same that let the Carboniferous scale trees reach high above the ground.

Plants have been using lignin in their cell walls for over 400 million years at this point. It was as big of an evolutionary unlock as bone was for our ancestors—something both strong and flexible, able to provide support but without being brittle. Lignin helped plants gain the structural support to start growing tall, not to mention forming bark and wood that would further open new evolutionary routes for land plants. Even leaves incorporated lignin, making them more resistant to the harshness of ultraviolet radiation, the acidity of rain, and the microbes and invertebrates that prefer to dine on plants. Lignin is so tough, in fact, that it continues its protective duties even after tumbling away from its parent tree, allowing fallen leaves to accumulate instead of immediately returning to the soil. In this way, even in death, trees changed life on Earth.

The spaces within the leaf litter aren't exactly warm in an absolute sense, but through the fall and winter the temperatures within the rotting blankets are warmer than the air above them. It's just enough that insect eggs, amphibians, and even chilly bats unable to find a warm cave can survive beneath. *Anancus*, sadly, is far too large to take advantage of the benefit. And while the leaf litter allows far more small animal species to survive in the forest year-round than otherwise would, the decaying leaves are also a promise to the new plant growth that will come up from the soil. The leaves refresh the soil as they break down, returning some of the vital organic compounds to the loam. The ground is replenished every season, the leaves' breakdown just as important to the plants that will come up in spring as the return of longer days. *Anancus* will be back then, when all the reds, yellows, and browns have once again fed curtains, canopies, and carpets of green.

15

After the Ice

15,000 years ago
New York State

SNAP. ANOTHER VIOLENT REPORT SOUNDS THROUGH THE woodland. It doesn't make birds in the nearby trees scatter, as the first had. For nearly an hour now, along the edges of a swamp starting to take on shades of orange and purple in the slow summer sunset, pops and cracks and the swish of boughs brandished through the air have sounded through the glade. They are the sounds of satisfaction, underscored by the low rumble of a shaggy elephant chewing entire conifer branches one at a time.

The mastodon isn't in any hurry. The shaggy beast has her pick of the more than three hundred plus pounds of vegetation needed for her daily meals. Still, she breaks off branch after branch greedily, chewing each one to pulp with her ridged molars larger than a human's fist. Her jaws have found their rhythm now, her burly and delicate trunk stuffing in morsel after morsel as she keeps grinding away.

There's not a hint of ice anywhere around the pond. This is summer, life all around her is awake and buzzing during the long and humid days. And more importantly, this is an interglacial. More than 7,000 years earlier this entire landscape was buried beneath two thousand feet of menacing ice. It was the ice that dictated life on half the continent, millions of square miles held beneath a scraping chill that carved mountains and drove great divots into the ground that melting ice would turn into lakes. Ice determined where life could thrive, what pathways to migration might be open and closed. Frozen water had not had such a grip on the planet in hundreds of millions of years.

The spread of the indomitable ice sheets has not been an unbroken freeze, though. It's not as if the Earth caught a chill it could not shake. The ice has come and gone through the millennia, a back-and-forth that has resulted in a split community of organisms—those that spread with the chill and those that thrive in the times in between. For two and a half million years ice sheets spread from the poles when the climate turned cold and dry only to retreat toward the poles as conditions fell back into warmer and wetter. Plants and animals moved according to what the ice determined. And while mammoths are creatures of the glacial periods, their bodies evolved to withstand the shocking cold while picking away at the willows and grasses of the frigid, exposed steppe, mastodons were always more comfortable in comparatively warm and muggy forests. *Mammut* overlap in time with the mammoths but truly thrive during the interglacial reprieves, following the forests and swamps as the ecological tide has gone in and out.

The hungry mastodon twists her trunk around another branch and pulls, bending the arboreal appendage until it pops away. She's so big that she doesn't jerk or lose balance with the tug. In one smooth motion her trunk dips below her tusks to deliver the fibrous food to her jaws. Those impressive teeth—one of the telltale traits that instantly distinguish her from a mammoth—have evolved for exactly this sort of task. While the molars of mammoths are relatively flat and ridged, great grinders for mashing low-growing plants into tiny pieces ready to go into their internal fermenters, mastodons will eventually be given their name because their molars are topped with paired, mammary-like cusps. The enamel-covered chompers have evolved over millions of years to bust up and grind through forests, the teeth of an animal that browses the trees rather than grazing endlessly on grass.

Even the way the great mastodon molars grow in the animal has been shaped by the plants the creature consumes. Mastodons—and mammoths, too, for that matter—don't grow new teeth like most other mammals do. The majority of mammals that have ever existed have been granted two sets of teeth during their lives. There are the milk teeth, the first set to develop, that are eventually pushed out of the way by a full complement of adult teeth. Even the great saber-toothed cats of this era develop teeth like this, going through an awkward phase when they have both their milk teeth and the barest tips of their adult fangs, giving their gangly forms an even more awkward expression. But during their 60 million years of evolutionary history, elephants have done something different.

Being big requires copious amounts of food. It's part of the classic trade-off—larger body mass means that you can eat a wider variety of foods, including food that doesn't have many calories in it, but you have to eat a *lot* of it. Plants haven't taken this in stride. They've evolved all sorts of responses to the ever-hungry mouths of herbivores, from caterpillars to giant ground sloths. Various plants have evolved tougher exteriors or forms abundant in silica, not so abrasive as to be immediately noticeable but enough to scrape away just a little more enamel with every chew. Ancestral elephants were faced with something of an evolutionary dilemma. If they got all their adult teeth at once, big elephants would grind down their teeth to useless nubs before being able to grow to full size. So the behemoths evolved a dental conveyor belt.

Our mastodon's adult teeth didn't come in all at once. Like most mastodons, she instead gets six sets of molars that slowly develop one after the other. Each one pushes the old, battered tooth out of the way as it comes in, providing a fresh chomping surface for the immense elephant. Naturally, even this adaptation has its limits. Elder mastodons sometimes chew slowly with worn-down and broken teeth from their last set, trying to run a biological equation of how fast they can feed versus their daily requirements. It's a common problem, especially because healthy adult mastodons are too big for any carnivore to safely hunt—disease, broken bones, infection, and accidental tumbles are more common causes of death once mastodons get to their adolescence. Even with an abundance of molars, mastodons have to make their toothy tools last.

Understanding the life of the forest without the mastodon is impossible, just as understanding the mastodon without the forest is a fruitless task. The woods are not just one great regenerative buffet for the elephantine beast. *Mammut* shape the forest with every step. And if the great beasts were to vanish, the absence would be felt among the landscapes where they wander from one stand of delicious trees to another.

From a tree's perspective, *Mammut* is a menace. Young saplings can be pulled out of the ground and chewed up in mere moments. Even adult trees aren't safe from the ravenous, warm-blooded appetites. Trees that grow among boulders, thorny shrubs, cliffsides, and other hard-to-reach places can persist through the years without being bothered by the big mammals, but those out in the open can be slowly pulled apart or toppled over. The big herbivores—mastodons, giant ground sloths, camels, horses, bison, and more—graze and browse at just about every level from the ground to the lower branches, each bite reducing the number of photosynthesizing leaves trees need to survive. A mouthful of leaves to a mastodon is an injury to a tree or shrub, requiring new growth that might be bitten off before it's even completed. Grasses may have initially spread during the past 20 million years due to climate shifts, but now the big herbivores maintain open spaces where woodlands are once again open and huddled together as if standing in defense of the next *Mammut* to wander by.

If *Mammut* and her megafaunal neighbors were to disappear, forests would once again close. They wouldn't be those of the Paleocene—warm, humid, and dense—but instead great tracts of shade-giving trees like maples, elms, chestnuts, and cherry that would press back against the expansive grasslands. Then again, some of the Pleistocene plants have survived by relying upon the big herbivores to ingest their seeds and carry them farther away. No single factor overwhelms all others, but represents a constantly unfolding relationship. When part of that relationship vanishes, the ecological dynamic can only change.

We'd be mistaken to think that the mastodon's role in the world is restricted to the forest, or that the journey of the boughs the elephant eats ends when they are mashed into woody pulp. Mastodons are hindgut fermenters. After their meals are mechanically and chemically broken down by both teeth and stomach, the cellulose-rich mash slips into the intestines. The longer that gastrointestinal trip takes, the more the mastodon can extract from the plants it eats. In the process, the mastodon leaves plenty of green pats behind it as it wanders through the forest, giving the ingested plants another life as food and nesting materials for beetles and other invertebrates. All that fermentation also means the mastodon farts as it trundles about, swinging its tail to spread the smelly, earthy scent. The Ice Age stink bombs are rich in methane, a gas that nudges Earth's climate.

Just as an individual plant doesn't photosynthesize enough to significantly alter the amount of oxygen in the Earth's atmosphere, no single *Mammut* or *Megatherium* squeaks out enough methane to change the climate. During this point in the interglacial, however, there are still vast populations of giant herbivores all over the world. More than a billion such herbivores are wandering the planet at this very moment, all of them munching, fermenting, and farting, which produces more methane than any other source. As much as 60 percent of the methane in the atmosphere is coming from megafaunal outgassing. It's keeping the world a little warmer, especially in the Northern Hemisphere where so many of these large-bodied animals roam. The legacy of the long relationship between the giant herbivores and their meals affects the world's climate, in turn regulating the individual lives of plants and animals around the world. If the megaherbivores were to disappear and stop converting trees into methane, among other things, the global temperature would dip.

Nature is not so strict as to fashion itself tightly around the actions of the fuzzy elephants, though, as if the mastodon were a dedicated gardener. Life has persisted for so very long because of flexibility. If the mastodon is to disappear, another large herbivore could step into the void and inadvertently perform many of the same duties. Trees and other plants have been growing alongside changing arrays of herbivores for millions of years. A mastodon trampling down seedlings, plucking up willow shoots, and shoving over trees is shaping the forests the same way that *Apatosaurus* did back in the Jurassic. Tighter relationships are harder to sever, as with some pollinating bats that are precisely adapted to the plants that rely on their help, but there is truly no such thing as perfectly adapted organisms. Such a creature would soon go extinct as the exact circumstances they were suited to changed, like the ancient brontotheres and long-lost apes that hugged the forest too tightly as the woodlands were cut back. Living things are always traveling across the edge of what they can possibly do and how they actually behave in their circumstances, the uncertainty of what might happen next meaning that evolution is sloppy and ever surprising. Mastodons are among the latest creatures to so directly shape the forests, but if the history of evolution is any guide, they surely won't be the last. Perhaps at some distant point, a million years from this moment, there will be giant herbivorous rodents or immense iguana descendants that take on the role of megaherbivore. As long as there are forests, there will be animals that mold their form.

Leafcutters are solitary bees that create bundles for their offspring from leaves and pollen.

It's easy to focus on big creatures like the mastodon because their behavior is so easy to directly see. All through the forest, plants are growing, being chewed on by insects, forming symbiotic relationships with fungi, photosynthesizing, and pumping out oxygen, among other things. The interactions are constant, even on the most still and sweltering summer day when all life seems to come to a halt. Every tree is an ecosystem in itself, every shoot growing out of the ground another life that alters the shape of the grove around it. It's a constant dance of making and unmaking, timescales long and short twisting around each other as one life becomes part of another. While the mastodons thread their way through the tree stands—and sometimes push over entire trees as they wander—much smaller lives intersect with all this vibrant growth, too.

The thin white trunks of a nearby birch stand tall as they slightly sway in the breeze. And on one particular leaf, on one particular birch, a small insect is busying herself with an essential task that must be carried out with the utmost care. She's a leafcutter bee, smaller than a fingernail. She doesn't belong to any hive. Leafcutters are solitary, minding their own business and creating small capsules for their offspring that do not require vast complexes of hexagonal cells. Her sharp mouthparts make quick work of the thin plant tissue. She comes at the leaf sideways, orienting the leaf's edge between her tiny jaws as she starts to bite, tilting her head down so that she cuts out a half circle from the edge. That's all she needs for the moment, just a snippet to attach to the others. Carefully holding the cutting in her jaws, she flies back to the rotting log where her babies should be safe for their vulnerable early days. Landing at the entrance of a hole made long ago by a wood-boring beetle, she wiggles in and positions the birch cutting against the leaf tatters she's already collected and adheres the pieces with her saliva. Not many more to go now. Once the last deposits are in place, she'll leave pollen and a dollop of nectar toward the back—food for the larva that'll emerge from the egg she'll lay in the cozy cell. If all goes well, a fuzzy little bee will cut her way out of the leafy burrito in about a month and begin a life of visiting flowers just as her mother had. Plants can be building materials just as they are food, and after so many millions of years of coevolution, the relationships of Earth's garden have continued to open even more complex interactions. In fact, there are even plants that have evolved to reclaim what other species have taken out of the trees.

Within earshot of where the *Mammut* is rumbling and turning tree branches into pulp, a trio of ghostly white flowers huddle together beneath the shade of an American chestnut. The small plants aren't bothered by the fact that the tree is soaking up the sun with its long, serrated leaves. They don't need the bright light. The food they rely on lives in the soil with them, the plant sipping on what grows in the dark. And just like the mastodon, they have only just recently returned to forests that have sprung up after the great ice sheets dragged themselves back toward the Arctic.

Photosynthesis predated plants by billions of years. Plants have only ever borrowed the ability, an accidental gift when ingestion turned into an ability to change components of Earth's abiotic makeup into living, growing tissue. The greening of the planet was so widespread that it opened opportunities for some plants to give up on photosynthesis altogether, instead taking a little from the great trees around them to thrive in the shade. *Monotropa uniflora* is such a plant, its single and downward-turned flower an expression of how plants had evolved to give up the very thing that has made them such influential parts of Earth's ecology.

Monotropa doesn't contain any chlorophyll. The plant has no means to photosynthesize. If it did, surely it would have to grow outside of the shifting shade beneath this chestnut tree. The plant gets its food from fungi, which in turn sample what the tree produces. The soil of this woodland—of any forest—is densely and thoroughly alive, crisscrossed with webs of fungal threads that gather food for the mushrooms and their relatives. *Monotropa* is too small and too singular to cozy up to tree roots alone, but it can skim from other living things. Clusters of the plant's tiny roots are covered with microscopic hairs that attach into the fungal filaments and draw in everything it needs to grow.

For a time, the plant retreated from these forests just as the mastodons and giant sloths did. *Monotropa* could not grow under a skyscraper's depth of solid ice. It was a journey that took place over generations, a journey of populations scattered on the wind rather than individuals shuffling to the south.

Even growing so low to the ground, hemmed in by so much else in the forest, *Monotropa* rely on the wind to reproduce. Wind or wandering insects can help members of different clusters pollinate each other, each solitary flower delicately dropping their fertilized seeds to the ground. Even a relatively gentle breeze can carry the seeds to new patches of the forest, and for hundreds of thousands of years luck has been with the little plants. The wind kept them one step ahead of the walls of ice to soils where the fungus they rely on still spread through the ground. New gusts nudged *Monotropa* back north as forests reseeded. Pushed out of their homes, *Monotropa* hung on and came back to twine themselves with the living forest floor.

Monotropa, as well as the leafcutters busily carving holes into leaves, will survive the coming changes. The mastodons and many other large creatures among these forests will not.

If all else remained equal, the relatives of the branch-breaking *Mammut* would persist for possibly millions of years more. But changes are already underway that will remake the world. Humans have been spreading around the world for millions of years, and the last surviving species arrived in North America more than ten thousand years before this *Mammut* moment. People have learned landscapes all over the world, often reshaping them with fire. Even though many plants have evolved to coexist with forest fires, the blazes can become so intense that the entire ecology of a place changes. Thousands of miles away, among the asphalt seeps of what will eventually be known as Southern California, woodlands where sabercats lap at streams and giant sloths pull branches down to their boxy faces first became dried out due to natural drought and then turned to ash by fire. The forest and much of its menagerie vanished, leaving a grassland populated by coyotes behind. And as people find their relationship with the landscape, changing it as they do so, the world is on the verge of a strange cold snap. About five hundred years from this moment, the warmer interglacial will be broken as Earth falls into a relatively brief cold snap, rapidly shifting to a freeze that recalls the days of the incredible ice sheets. A drop of about six degrees Fahrenheit might not seem like much, but it will be enough to knock entire ecologies off balance during a time when populations of many large mammals—including the mastodons—have become fragmented from surviving round after round of glacial expansion and contraction. Habitats will become inhospitable to their large herbivores, who then can't provide food to *Smilodon* and the dire wolves that have evolved to specialize on large prey, sparking a collapse that will affect species large and small. Even tiny creatures, such as cowbirds that hop among the grasses where giant plant-eaters browsed to eat the bugs they kicked up, will perish.

The mastodon browsing along the edge of the swamp will not see these changes. Her time is a split second against changes that play out over incredible time spans. But she, the leafcutter bee, and *Monotropa* are nevertheless part of the same story. Vast changes are often made up of tiny interactions, the smallest choices and happenstances that can reverberate and build to change the very nature of the landscape. Every creature carries a legacy of the past as it walks toward a future it will never see, an ecosystem in itself and part of webs too large to perceive. Plant and animal shade into each other, shape each other, and bring each other into existence through nothing more or less than an ongoing relationship. What might grow tomorrow, only tomorrow can tell.

Conclusion

THE PRESENT IS ROOTED IN THE PAST JUST AS A TREE IS rooted to its home soil. There are places on our planet where we can follow the dendrology into chronology, in touch with a time we missed but can still feel through life's unfolding continuity. I don't just mean mastodon bones or the remnants of leaves in Ice Age sediment held in museum drawers. There still lives an organism who witnessed how the world changed as the glaciers receded, a survivor whose life makes mine seem like nothing more than a fleeting ember wafting on a breeze. All those hundreds of millions of years we've watched unfold have brought us to the present just as surely as if we had followed a branch from its junction with a tree trunk to the budding tip of a branch.

Pando was nearly tucked in for the winter by the time I'd arrived. My partner, Splash, and I, our German shepherd Jet in the backseat, had been hoping to view a terrestrial sea of yellow-and-orange leaves shifting and shaking in the early autumn breeze, a good excuse for a road trip to a part of Utah we often speed right by. But my timing was off. We'd come weeks too late. Relatively few parched leaves clung to the aspens as we rolled along the county road that interrupts the "trembling giant." When we hopped out of the car to walk among the pale poles, the sound of the heart-shaped leaves was less the summertime applause I associated with the tree and more of a dried-out rasp no doubt caused by the record-breaking heat of yet another Beehive State summer. Still, a tree sliding into bed for winter is no less splendid than it is on the longest day of the year, especially when that tree is arguably the largest living thing on Earth.

With Jet twitching to his doggy dreams in the backseat, Splash and I had traveled almost three hours from Salt Lake City to see Pando. It felt strange to have such an exceptional plant growing a day trip away, standing since the last Ice Age in what's now delineated by the US Forest Service as Fishlake National Forest. You really can't miss the tree. There's a small sign that declares ENTERING *the Pando Aspen Clone* as you follow the county asphalt over the hills, the *LEAVING* sign less than a song's length down the road. The true extent of Pando spreads much farther from what you can see outside your windshield, not so much a forest but a single entity made up of many individual parts.

Perhaps other clusters of aspen, smaller huddles of clones festooned with heart-shaped leaves, might have sensed that Pando is something different. Our species, however, requires signage. Pando doesn't look like a distinct individual to most of us, save for the few who've interrogated the forest to find the edges of the tree. The trappings of Western culture haven't done us many favors, either. Many of us are used to thinking in individualistic terms—there is one aspen tree, next to another, next to another, the same way everyone standing in concert line is individually distinct. But human life is a poor measure for the experience of any other being. We need organisms like Pando to expand our understanding of what a life can look like, the connections both hidden and obvious where distinct entities create community.

Even though Pando has been estimated to be as old as 14,000 years, when mastodons and giant camels might have trod the same woodlands, we've only known of the tree's existence since 1976. Back then, the broad span of clones were recognized through aerial photographs and on-the-ground comparisons of the tree's bark, leaves, and stems. More recent genetic studies found that Pando covers even more of its home basin than the initial surveys suspected, a plant that grows over 43 hectares—more than three times the size of every professional American football field put together. And as wispy as aspen trees can look, if you were to consider all of Pando together the entire tree would weigh over 6,600 tons. That's the equivalent of 1,100 adult American mastodon, herds and herds of the herbivores to match one aspen.

Listening to the crinkle and swish of the leaves underfoot, Jet happily panting as he set about his favorite challenge of sniffing the entire forest, I walked a short way into the grove and admired the aspen eyes peering back at me. I could look in any direction and start counting each of the pale poles, but none were truly solitary. All were expressions of a whole, part of a life that touches a time I can only know through fossils and boulders left behind by fleeing sheets of ice. The aspens started to feel like the ribs of some immense being that I was sunken down into, its tissues and systems surrounding me as I gazed up at the sky like a tick on a dog's back. It reminded me of one of my favorite dioramas I visited at the American Museum of Natural History while growing up, leaf litter and various invertebrates enlarged to make me feel tiny by comparison, the same feeling I got standing next to the skeletons of *Brontosaurus* and friends elsewhere in the marble halls. Some wonders of nature are acquired tastes, requiring changing how we think or perceive the world, but some expressions of nature are so grand that they force us to think of ourselves in another context. Pando is one such prompt among many.

Running my fingertips over the bark of one trunk, I tried to think through the rings of that one part's life and whether the decomposed, recycled remnants of previous Pando parts had helped feed its growth over the tree's expansive expression. Even in that moment of oncoming dormancy resembling death on that October afternoon, all the scattered leaves Splash, Jet, and I walked over formed a haven for tiny spiders, insects, and other creatures that overwinter beneath the shielding cellulose just as they did back in the days of *Anancus*, and probably more ancient times still. How many lives can one life touch? How many living things have alighted on, chewed up, dwelled within, pushed over, and otherwise had a brush with a tree so enduring it probably understands the nature of time better than I ever will?

Of course, the forest I watched my genetically modified wolf gambol through was not the same as would have been seen by *Mammut* or the Indigenous peoples who dwelled here during the megafaunal heyday. Pando is a dendrological ship of Theseus, each tremulous pole growing where others have no doubt poked out of the soil, reached toward the sun, and been folded back into the soil's warmth over time spans long enough that it's impossible to find sufficient depth of empathy to understand what such a life must be like. I didn't come to simply pay my respects to an old-timer, though. I wanted to take the presumptuous move of sitting within the great botanical thing so that I might think about what it might be like to live in such a different shape, an entity that is both its own self and community all at once. I drove all the way out to see Pando because I needed some way to visualize hope through the cold ahead. Standing within a living thing that's both ancient and a momentary embodiment of a story billions of years old, I came looking for hope.

Emily Dickinson famously wrote that "hope is the thing with feathers." The flutter of magpies and the croak of ravens soaring above burnt orange sandstone. But hope is not strictly the domain of the dinosaurs. Hope scatters itself on the wind. Hope buries its seeds for the next growing season, or perhaps the one after that. Hope grows, even pushing its way through concrete and chain link if it has to. Plants are both fulfillment and the beginnings of new hopes, seedlings that might not all make it but know nothing else but reaching toward the sunlight.

A feathery hope might seem like a fleeting thing, as ephemeral as a breeze. The hope of plants, however, is hope plus time. It's hope that has changed the world from the nature of the air to the blooms we give each other in both celebration and mourning. Through and through, plants embody hope that cannot be suppressed and has no choice but to try to thrive, even in the most hostile of places.

We often speak of hope as a fragile and delicate thing, almost as if it's made of glass. It can seem as thin as belief, an expectation of something we can't see yet and may never see. But hope is also something we do, a facet of our nature. The same is true for plants, even though they cannot entangle themselves in the abstractions we invent. We meet plants at different moments of becoming, ancient cycles that consistently bring about tomorrow and the day after when life will be much the same but always a little bit different. It's these rhythms that oxygenated the atmosphere, caused forests to take root in rocky soils, and make deserts bloom following the rain. Plants can do nothing else. Without the hope of a favorable wind, a break in the drought, or seeds finding just the right patch of forest floor, nothing about them makes sense.

I came to Pando with hopes of my own. Standing within the grove, within another living thing that might not even perceive my presence, I came with the hope that the lives of such organisms might help me better express the new roots I'd sent down since I found new soil of my own. I'd visited Pando both in my middle age and as a young woman, a still-growing thing after more than three decades of relative dormancy before embracing the truth of my transgender self. I'd been misidentified, misclassified, for most of my life, "like a lock that doesn't turn, like a plant that doesn't grow," as one of my favorite somber songs by The Menzingers put it. Coming out and beginning my transition shifted my perception as well as my body, making me wonder what else I missed during the static of those pre-transition years. Neophytes are never guaranteed fertile soil or a welcoming habitat, however. I'd started to bloom at a time when people like me would be considered weeds in what some wish to till as a godly garden, as if we were noxious aberrations to be uprooted. We are facts of nature as much as an aspen grove is, part of the broader expanse of our species, yet we're singled out as so other that some around us cannot see our humanity. I came to Pando to sit with the giant and reflect, to think on how such a survivor has persisted through all the uncertainties that eventually stack up as history.

I looked to plants like Pando exactly because they are so different from us. More than that, they have existed for longer than any of us can comprehend and are so ubiquitous we often don't think of them apart from the scenery. It's not challenging to find how their nature intertwines with queerness, itself a core part of all natures—including ours. We might think of queerness as what's bizarre and separate just by virtue of its name, but we know from ourselves, and nature itself, that the queer has always been part of nature's chaotic exuberance.

We don't need to go far to see how queer plants are. A crab apple tree growing in the yard or the roses we give each other to celebrate are plants that contain marks of what we may deem both male and female in the same plant, betraying our bias to split these identities even when they are very much part of the same organism. The arrangement is known among botanists as perfect flowers. Of course, there are plants in which what we designate as male and female parts exist on different individual plants, or different parts of the same plant, or involve varying expressions over a plant's life, an entire array of different arrangements that could hardly be called basic. And all these different ways of existing and reproducing provide the raw variation which capricious forces such as natural selection act upon, diversity feeding into new forms that will then form the basis for what comes next. And just as an individual plant may express different sex-related characteristics in varying parts of its body or at different times of its life, the same is true for our species. I know because I've lived it, watching over the years as shifting hormone levels have shaped my body. Every body, yours included, contains the capacity to express in ways we might think of as male or female in different combinations, all depending upon our life histories and the choices we make for ourselves. We can consciously do what plants do just by existing, a constellation of variable combinations that can alter how we meet the world.

Such diversity brings about even more diversity through time, not just along genealogical lines but from their home soil to the atmosphere itself. Through hundreds of millions of years, we've seen how plants have changed the world simply by existing. Perhaps more than any other form of life, plants have shaped nature on Earth and will surely continue to do so when we're fossils ourselves. We love and celebrate their diversity, not just the pretty flowers we're attracted to as much as nectar-hungry bees, but in the full panorama of the forest—the ferns, mosses, trees, epiphytes, and more that create what we try to summarize in a single word, a somewhat static expression of a living and ever-shifting community. I can't help but feel the same about our burgeoning and irrepressible queer community. We are not just a field of pretty wildflowers, spreading pollen like glitter in June, but a tangled wood overflowing with different forms and behaviors that only become more varied as we continue to explore this part of our own nature that has been repressed in the Western world for over two centuries. We are expressing a fundamental part of what life does, what life must do, outside of sterile ideological boxes that only serve to turn us away from the complexity and unabashed beauty of *life*.

At the conclusion of another book, published more than a century and a half ago, Charles Darwin likened life's shape not to a single tree but to "an entangled bank." Such a setting is too complicated to draw into neat diagrams of relationships. A tree of life was easier to understand, each branch neatly ordered even as biologists continue to prune and graft the precise arrangements. I think Darwin was onto a more fundamental truth by envisioning a moist span of mosses and grasses, clambered over by insects, lives so intertwined that they'd seem inextricable from each other. Perhaps that's why it's such a perfect encapsulation of not only how life evolves—through interaction and relationships—but also how we might think of queerness. Signs of the deep past grow next to, beneath, and upon what's novel, varied species on an entangled bank, a forest, or a desert for that matter, individual lives constantly bringing those habitats into being. Among these places we can find history as easily as what's new and ephemeral. In a forest we can find liverworts that resemble the earliest land plants growing near towering, brushy conifers not unlike those that fed some of the largest creatures to walk the Earth, mingling with deciduous trees whose ancestors managed to grow back faster in the wake of the catastrophic extinction that marked the last days of the non-avian dinosaurs. All of these expressions are part of the word "forest," just as "queer" encompasses an ever-expanding community of different expressions that carry echoes of their history with them. Both forest and queer are not flat descriptors, but terms that feel alive—states of being that have a past, present, and future, never quite the same today as they were yesterday.

The world we meet today is the foundation for tomorrow. If we are the plants, then our ideas and actions are the seeds and spores. We might not live to see them grow or even flower as we have. Nevertheless, the variety we embody and bring to the world forms a base from which new communities will grow and respond to the shifting world around them. There's really no telling what might happen next. Change can grind or come in a shocking instant. Through all of life's changes, though, plants have responded, adapted, and thrived anew, changing the world every time they flourish.

And still we grow toward the light. Just like plants, it's something inherent to us that we cannot help but do as part of our existence. Anyone who has had to navigate their own path to coming out knows the feeling that they are bending toward something, growing toward a point they have not reached just yet. Even when we do, everything we learned through introspection meets the world and we continue to change and grow, because how could life do otherwise? Even in moments when we feel ourselves cast into the shade, even uprooted from our home soil, we still grow toward the warmth and feel our roots twine with those of others who know these feelings.

We aren't guaranteed a world without barriers, fences, or instruments meant to cut us back, but, within our queer forest, we will always grow back stronger the next year. Even the faintest and most fragile new shoot is a wonder that will change the world around it.

Whatever experiences of life they have must be fundamentally other than our own, so much so that we have barely begun to understand how plants perceive the same world we inhabit. We share foundational aspects of our biology from billions of years back and yet have come to arrive at the same time so different from each other. Not fussed with our opinions, they reach and wind and lash and above all *grow*, changing the world as they do so. Considering them and what their lives might be like requires that we broaden how we think, acknowledging how limited our personal experience might be, and I can only say that I am in awe of how beautifully and lavishly nature expresses itself. My sense that I'm part of this variety, a creature full of possibility, stems from this view of nature, where any single palm, aloe plant, cactus, or blade of grass is but one expression of what's possible among untold lifetimes and lineages. It's an ongoing practice in empathy, to notice a life so different from mine and wonder, "What is life like for you?" even if no response will ever come. Life is messy, strange, persistent, and more varied than we could possibly imagine. We can't grasp the full nature of the idea, but we can at least approach it and, perhaps, begin to acknowledge that we're creatures of possibility, too.

Appendices

WHAT IS A TREE? I KNOW THIS SOUNDS LIKE THE KIND OF question inspired by a different kind of plant, like the catnip our *Machairodus* friends enjoyed back in the Miocene, but it's still one worth asking.

Trees often seem obvious, especially to those of us who dwell in urban habitats where parks departments intentionally plant maple, elm, and even stinky Bradford pears along streets and greenways to make us feel a little more at home. But under the broader canopy of botany, a tree is something like a martini—not something defined by its contents but by its overall shape. The first trees, like the great scale trees of the Carboniferous, were plants most closely related to species that only persist low to the ground, huddled in the undergrowth, today. We call them trees because they grew tall and cast shade over sucking swamps where amphibians lazed and bird-sized dragonflies buzzed, not because they have much relation to the dogwood in your yard. And even now, there are some plants such as juniper that blur the line between tree, bush, and shrub, plants that have a vaguely tree-like shape but spread nearer the soil, evolutionarily close to something we'd define as a tree but isn't. Through an evolutionary lens, the word "tree" is like "fish" in that it makes complete sense until it doesn't—an entire cast of seeming outliers telling us that our definition was founded on sand. We can perceive nature, but the way we speak of it and understand it will always be a matter of our own invention.

The fossil record doesn't make such demarcations any easier. Fossil plants are rarely preserved in their totality, root to leaf. More often, if not most often, we get little more than a chunk of trunk, a leaf, a gnarled root fragment, the scattered pollen that settled in the stone instead of alighting on its intended partner. Many of these odds and ends get names, following scientific protocols of nomenclature put forward over two centuries ago. We feel unmoored without labels, and so every prehistoric plant goes in its pot. But taxonomy is so easy to snarl, and paleobotanists may inadvertently give each individual part of a plant different names until a more complete specimen turns up. At that point the systematists set about pruning the list of names, cutting the more recent titles until the first and most unassailable title remains. Despite the work, paleontologists often feel relief at these revelations. The scattered puzzle pieces have become something new, another pinpoint to tie threads between and draw out from.

Of course the incompleteness of the fossil record is upsetting. Every new fossil is a reminder of our good fortune that we know anything at all and carries grief at what we know we'll never uncover. There are entire environments that have vanished as if they were never there, flourishing for a brief moment with nothing at all to preserve them. And of course we could focus on the incompleteness of the fossil record as a problem to be solved or confounding factor to be conquered as many paleontologists have done ever since nineteenth-century naturalists like Charles Lyell and Charles Darwin wrote apologia for why their Earth-moving ideas on the past did not yet have extensive fossil records to prove how life changed. In all the time I've spent grinding sand into my socks and having my blood lapped up by desert gnats, though, I've come to appreciate the missing pieces. We can't understand the fossil record without gaps.

Suppose for a moment that the geology of Earth provided us with a much more complete record. It would feel alien to us. Instead of the stratified brain teasers Earth scientists keep picking away at, there might be onion-like layers of each sequential age that included a full and complete accounting of every population of organisms that has ever existed. (I would say "every organism that has ever lived," but, aside from other considerations, animals have to eat, and asking for pristine representations of each meal would be a little too much to ask even in my wildest paleontological dreams.) With such a fantastically complete record we could start from near-yesterday and pick our way backward, tracing lines of descent and following how anatomical adaptations became embodied in different species. So many mysteries might be solved. And yet such a record would require an intense rethink of biological fundamentals, including how to differentiate one species from another. We only tend to notice the origin of a new species when we can perceive something *different*, in bones, genes, or something else. There's a shift to something new that seems consistent, even for a limited time. But in a complete record, recognizing different species would become an extremely fraught task—the subjective organizing systems we've invented would cause an incredible number of migraines in trying to disentangle what's shared from what's different. In other words, the fossil record would make us have to tangle with how messy nature truly is and we'd have to invent a new way of thinking about nature to even begin drawing out bigger conclusions.

The planet we live on creates fossils by pure happenstance, not assurance. Gaps simply come with. The empty spaces, and wondering what might exist between them, stokes our curiosity and leads to all those pleasant moments when we know we're wrong.

I selected the vignettes that fill this book because they represent significant moments in the long relationship between plants and animals. Some of these moments were small shifts with enormous implications, such as when plants pumped incredible amounts of oxygen into the air or the spread of grasslands in a cooler, drier global climate. Others are smaller stories that highlight some of the strange ways plants and animals have evolved together, such as the happy sabercats and their catnip or ginkgo trees evolving ways to survive the appetites of giant dinosaurs. Time and again, plants would do something new and then other forms of life would shift in response and set up the next change. And I was able to select these particular stories because of the fossil record's imperfection. Each story relates to times and places that are somehow exceptional against the fossil record, spots in time when we can get a sense of call and response, intertwining to produce relationships and shapes that would have previously been inconceivable. In a totally complete record, little would stand out as special and I could snip out almost any ancient setting to make a point. In a flawed history, when we are left with hanging questions, we can more easily find our sense of curiosity.

The narratives in this book have all grown from the technical literature. Of course I've added some embellishments and assumptions along the way. It's unavoidable. The most scientifically accurate book in the world would be little more than a ream of data that would be the death of wonder. The point of spending all this time flipping through technical research papers, walking museum halls, and dripping sweat all over the desert is to envision what all these long-lost friends of ours were like in life. Any restoration of a life long gone carries the inherent risk of being mistaken or missing some piece of as-yet-uncovered information. When I was more of a sapling, everyone knew that the huge insects of the Carboniferous got so big because of increased oxygen in the ancient air that allowed them to breathe more efficiently. Now we know differently, with the surplus oxygen being a poison that arthropods had to cope with by being larger in their larval phases. Every nonfiction author worries that something about their story might be contradicted in the time between sending off the final draft and publication, and with new paleontology studies appearing faster than it's possible to read them, all I can really do is make peace with the fact that I am not an unimpeachable interpreter of the past—no one truly is.

Nevertheless, fossil visions grow from the ideas and findings that have already been dusted off. Each of the stories in this book has multiple stories behind it—from the wonder of discovery to the ethics of how we study the past. In the spirit of that grade-school exhortation to "show your work," I've included this appendix to provide some basic background on what we know, what I've reconstructed, and how I've shifted focus through each chapter. My hope is that each of these short sections will feel less like an explanation and more like a series of threads you could begin pulling at yourself, if you so choose, a branching canopy of possibilities that can help you perceive the past in our ever-shifting present.

Appendix I

Of all the chapters in this book, this first one is the most likely to change.

You know the phrase "like looking for a needle in a haystack"? That's nothing compared to the quest to find evidence of life from the first three billion years of its existence on the planet. It took that long for organisms we can see with the naked eye to be preserved in the fossil record, itself an incomplete collection of ancient life. I know from experience how difficult it can be to find the remains of an eighty-foot dinosaur in rocks where you know they must be. Imagine the challenge of uncovering the bacterial and microscopic life that can document some of the critical evolutionary moments when life on Earth became what it was. It's a wonder we have any understanding of life's first three billion years at all.

The search for early life is full of controversy. Sometimes we don't find life itself but its effects, such as iron-banded formations related to the oxygenation of the planet spurred by cyanobacteria. Other times paleontologists go back and forth over what seem to be cells detected with powerful microscopes, whether the specimens in question really represent life or are irrelevant pseudofossils. And keep in mind that all of these finds are coming from incredibly old rocks that are relatively rare on Earth's surface, slivers of time in the process of being ground away by erosion. The earliest parts of Earth's history are being destroyed by the planet's own processes. That we can understand anything about Earth's early life is a testament to the persistence of generations of geoscientists. As much as we've learned, however, there is always the possibility that someone announces a discovery that shifts our ideas about life's formative years. For this chapter, then, I attempted to center on fossils that are relevant to the rest of the book's story even as new contenders for older organisms are likely to be announced.

Overwhelmingly, the story of life on Earth can be told as a cellular tale. We're newcomers and anomalies, multicellular beings made of various tissues, organs, and systems jangling around, single-celled life living on us and within us in a world where such organisms have proliferated for over seven times longer than there's been animal life. And that is precisely why the origin of the very first plants is so difficult to study.

The history of the earliest organisms capable of photosynthesis is a little easier to approach. Studies of living cells that don't require oxygen to photosynthesize have given paleontologists a few telltale clues to look for, and experts suspect that anaerobic photosynthesis came long before the sort that filled the oceans and atmosphere with increased oxygen. The shift to oxygen-focused photosynthesizers is visible in the rock record, as well. That's because photosynthesis altered the composition of the oceans and the air by dumping incredible amounts of oxygen into the world. Geologists and paleobiologists often look to stromatolites and banded iron formations—alternating stacks of vibrant red and black stone—to detect when the organically created surplus of oxygen began to change the planet. Cyanobacteria are the most likely culprits, organisms that still thrive in both fresh and salt water and, if we allow ourselves a little poetic license, can be said to have seen the rearrangement of continents and the rise and fall of entire groups of other organisms.

Plants evolved after such photosynthesizers changed the world, and so their signature on the planet is swamped against the background. If we're going to identify the first plants, then, it has to be on the basis of anatomy and what such an organism would share in common with all living plants—from pine trees to turtle grass—that we see around us today.

Paleontologists have identified multiple candidates for the earliest potential plant. They are all extremely tiny, so small that it is sometimes difficult to tell whether their cellular features were parts of the living organism or are artifacts of preservation. In 2017, for example, paleontologists proposed that they had found the earliest known plant cells in the 1.6-billion-year-old rock of India. Some of the fossils seem to have pit plugs, or holes between two algae cells only found in red algae, which would identify them as the oldest plant forerunners yet found. But some experts doubt that the pit plugs are the real thing, meaning that the fossils could be from some other form of photosynthetic life. There are even older possible plants from Michigan in the United States, Shanxi Province in China, the Congo Basin, and elsewhere, but all of these occurrences have been questioned for one reason or another. Given that, I chose to focus on the earliest organism paleontologists agree is relevant to plant origins— *Bangiomorpha pubescens*.

Described for the first time in 2016, the *Bangiomorpha* fossils are tiny—about two microns across. Their name is a reference to modern red algae, *Bangia*, and, in what's certainly a unique choice, the resemblance of the filaments to pubic hair as a reference to the plant's relevance to the evolution of sex. A stack of them wouldn't reach any taller than about two millimeters, about the width of a thin wedding band. Comparing the living plant to the fossil, the red algae that grows in saltwater shallows is a whisper of how plant life started over a billion years ago.

Bangiomorpha probably wasn't just like an ancient copy of its living counterpart. Paleontologists have yet to find representatives of all parts of the plant's life cycle, and it seems *Bangiomorpha* grew in a slightly different way. Nevertheless, fossils of the plants show traits that correspond to the sexual differentiation of red algae cells. The ancient plant was capable of differentiating its cells and reproducing with spores rather than simply splitting. The fact that *Bangiomorpha* was fastened to its substrate by a holdfast, or anchor, is also critical because it's an indication of a cell taking on a specific function and therefore multicellularity. It's likely that organisms that sexually reproduced and had multiple cell types in the same body go back even earlier, but *Bangiomorpha* is the best evidence paleontologists have yet uncovered.

Appendix II

Finding the first *anything* in the fossil record is a fraught task. Maybe that's why we obsess over the superlative so much. We construct our image of history according to critical personas, places, and dates, and the fossil record is no different. The first dinosaur, the first human, and, in this case, the first land plants all become focal points because we want to get to the roots of what we so often cast as great dynasties. And, of course, it's an impossible task.

Often, when we speak of the first member of an evolutionary lineage we're talking in terms of family relationships. The task is fundamentally fraught by the fact that—for example—what might be considered the first mammal would not be very different from the almost-mammal that preceded it. This is another instance in which gaps in the fossil record work with how we organize the past, with breaks in the record allowing us to understand a particular species as closer to this or that part of a spectrum. Paleontologists define these breaks on the basis of shared characteristics, but in a complete record the differences might be so minor—even imperceptible from fossilized remains—that it would be a matter of splitting hairs. At least, in the case of the first land plants, we are focused on what the greenery was doing rather than what family it belonged to.

We still remain at the mercy of an imperfect record. Experts can certainly pinpoint the earliest land plant that we know about, but not the first there ever was. To do so would require a global fossil record pinpointed to a timeline we do not possess, so we're unable to say whether the first terrestrial plant took root on an ancient Monday or Friday, during a Northern Hemisphere spring or an austral autumn. We have to make peace with the fact that for any "earliest" contender we find, there was likely some living thing that was there even marginally earlier that was swept away without even a whisper of a hope of being included in the fossil record.

I had hoped to include more in this book about the early history of plants in the seas. After all, oceanic plants—

from turtle grass to algae—are extremely important parts of Earth's ecosystems today, both locally and globally. Most of the oxygen we breathe isn't coming from the trees in the local park but from algae out in the seas. Unfortunately, we know relatively little about what happened between the time of *Bangiomorpha* and the shoreline liverworts in this chapter. As you might expect, ocean plants would be squishy, easy to decompose, and difficult to positively identify. Many histories of plants put the emphasis on when plant life arrived on land, associating the word "plant" with the terrestrial realm even though we know many plants live in watery environments, too. Perhaps it won't make headlines the same way a new dinosaur species will, but there is a great deal left to discover about the early history of plants in the seas and on Earth in general.

The precise identity of Earth's first land plants isn't entirely clear. The best physical fossil evidence we have are only fragments of plants. Those tiny pieces, found in Oman, resemble modern liverworts—broadly considered to be primitive plants. Research since that discovery has suggested that plants began to grow onshore even earlier, changes to oxygen and carbon cycling in the rock record hinting at the presence of plants by about 520 million years ago—around the time that animal life in the seas was in the middle of the Cambrian explosion.

Lacking an extensive fossil record, what we presently think about the earliest land plants comes from two sources. The first is the liverworts, mosses, and similar plants that grow low to the ground, rely on some connection to water for reproduction, and live in niches similar to what we expect for the first land plants. Experiments with mosses, for example, have suggested that early land plants weathered minerals like phosphorous out of Earth's rocks in greater quantities than abiotic forces like wind and rain ever had, changing the geochemical composition of the sea as well as terrestrial habitats. The assumption here is that the earliest mosses and liverworts lived in a very similar way to their modern counterparts, more than 450 million years on, which will be difficult to investigate further without better fossils.

The second category of evidence comes from geochemical studies that consider the way plants affect the planet. We know from our current climate mistakes that burying plants is a way to sequester carbon and reduce the amount of carbon dioxide in the atmosphere. Earth's geochemical record seems consistent with the spread of land plants reducing carbon in the atmosphere—taking it in to build their tissues—while increasing the amount of oxygen in the air both through producing oxygen through photosynthesis as well as being buried quickly enough that plants don't decay. The relationships between plants, rock weathering, climate, and atmospheric composition will likely change as geologists learn more. These are complex global systems with many different facets to them. Still, it's not a stretch to see how living things that build themselves with carbon, produce oxygen, and eat away at the rocks they grow among could lead to major changes to the planet. If plants had not come ashore, or had come ashore at a different time and therefore with a different set of abilities, the entire history of terrestrial life could have been very different.

For the purposes of this story, plants set the conditions for big changes by providing a reason for animals to follow onto land. If we were to trace back through our own prehistory to those ancestors who lived between the water and the wet shore, we'd find fishy creatures that were munching on the arthropods that had themselves emerged from the water to feed on plants in a world relatively free of predators. Without those crunchy morsels our ancestors would have likely remained in the water, and the arthropods would never have emerged without crisp greens to tear apart with their complex mouthparts. Rather than being a major evolutionary triumph that many experts have called an "invasion" or "colonization" of land, our ancestors were merely continuing an ecological transition that was well underway and would have continued to alter the planet with or without them. It was plants, not fleshy-finned fish, that changed the world when they came ashore.

Appendix III

We live in a world directly connected to the Carboniferous. The period's designated name is the critical clue. So much of the coal that powered the belching factories and refineries of the Industrial Revolution to today came from great seams that formed back in the days of enormous arthropods and ancestors of ours who had only just begun to become accustomed to life on land.

When I look at artistic renderings of Carboniferous forests, I can feel the humidity drive sweat down the back of my neck. The ancient groves are the epitome of primordial. They almost look like an overgrown version of a terrarium, everything far larger than it would otherwise be. But that's largely because forests were new, plants having only just evolved the necessary tissues to reach just that much closer to the sun.

Anyone who has ever grown tomatoes in the garden knows that plants cannot grow infinitely tall on their own.

There are no Jack and the Beanstalk vines that can twist and twine to the clouds just through continual growth. Just as our bones help hold us up and strengthen the core of our bodies, so did Carboniferous plants evolve a compound called lignin that would be critical to the history of life on Earth—and some of the damage to it we must hold ourselves accountable for.

Lignin, in short, is what allowed plants to start growing up. It's a molecule made out of carbon that makes the cell walls of plants tougher, providing resistance against damage from the outside but also adding structural support. And while lignin was previously believed to be something unique to land plants and their closest relatives, recent research has found that much, much older photosynthesizers were capable of making lignin, too. Genetic research has found that red algae, encased algae called diatoms, microorganisms called dinoflagellates, and other unicellular life have long been capable of making the single links, or monomers, that lignin polymers are made out of. No one is entirely sure why, although it seems clear that these aquatic cells were not using lignin to grow impressive statures. Instead, researchers suspect that lignin might have offered some aid in defense from other cells or perhaps even provided some protection from UV radiation in sunlight—a kind of natural sunscreen that would have an advantage for living things that rely on the sun to make their food. It was only after plants started to grow on land that there would be a structural advantage to lignin monomers chaining together to make proper lignin polymers, making plants tougher against damage from other living things, offering more structural support, and assisting with water transport in some of the earliest trees. The success of lignin would also have lasting effects on the entire planet, especially as we now burn what remains of plants that reached for the sky thanks to the compound.

If I had written this book a few years ago, I would have stressed the importance of lignin for the preservation of the scale trees and other plants in coal swamps. The story has become more complex in recent years. When I was a college student, what unfolded seemed very simple. Plants began to grow tall because they incorporated larger amounts of lignin in their tissues, strengthening their trunks as they grew more than one hundred feet in the air. Lignin was such a tough polymer that bacteria and fungi hadn't yet evolved to break the substance down, meaning that downed trees tended to sit in the coal swamps long enough that they could be buried. This was supposed to account for the vast amounts of Carboniferous coal preserved around the world. Plants were also credited with the evolution of gigantic arthropods and amphibians during this time. Greater oxygen levels made breathing more efficient, allowing bigger animals to more easily get what their tissues needed. But now we know that what unfolded was not so neat and clean.

Lignin, or at least its precursors, was made by plants when they still lived in the water. Plants didn't invent lignin to grow taller. It was a compound that already existed and was put to a new use. On top of that, paleobotanists have found evidence of fungus damage on Carboniferous trees. The finds indicate that there wasn't any evolutionary lag between trees and their decomposers. Bacteria and fungus were perfectly capable of breaking down dead trees. The increased amount of plant material that became buried in coal swamps had more to do with Carboniferous conditions being so favorable for plants that mosses, quillworts, and horsetails were growing

faster than the dead ones were decomposing. The Carboniferous coal seams that powered the Industrial Revolution and
kicked off Earth's human-made climate crisis were preserved
because an incredible amount of plant life was growing in
places where it was likely to be preserved, subsumed into the
Earth's crust and later pushed back up to the surface through
mountain-building and the slow shifts of plate tectonics.

The insect story isn't quite so simple as it used to be, either. The giant dragonflies of the Carboniferous, like *Mega-
neura*, have been icons of the age ever since their discovery. But
dragonflies start off as aquatic larvae, as do many other "giant" insects of both the Carboniferous and other time periods.
Studies on modern larvae of such insects indicate that they are
very sensitive to increases in oxygen comparable to what happened in the Carboniferous. The evolutionary pressure favored
larger larvae, less vulnerable to excess oxygen, and opened the
possibility of larger adults that were better equipped to withstand the higher levels of oxygen in the air. The fact that it's
often insects with water-dwelling larvae that tend to get large
in times of higher oxygen, but not other insects, hints that conditions in the water were critical for these animals and coping
with too much oxygen was more important than any benefit
oxygen might have had for adult arthropods.

I was also fortunate enough to learn about new fossil discoveries while I was in the process of writing this book, including new information about some of Earth's first herbivorous
vertebrates. To give credit where credit is due, invertebrates
had been munching on terrestrial plants for millions of years
before the Carboniferous. The earliest land plants, in fact,
likely attracted the more amphibiously able invertebrates to
crawl out onto the shore and start nibbling at food sources that

their underwater neighbors couldn't reach. These exoskeletal morsels, paleontologists suspect, acted as live bait for fleshy-finned fish that could extract oxygen from the air as well as from the water, beginning a food web that eventually allowed our distant four-legged ancestors to slide their soft bellies over the Silurian sand. It would be millions of years more before the descendants of these fantastic fish would feel more at home on land, with arms and legs capable of carrying them over the land and the evolution of the amniotic egg—a shell-cased capsule that could allow embryos to develop in their own private ponds even at a distance from the water's edge.

Plants were already thriving on land by the time vertebrates became capable of spending their whole lives out of the water, covering the rocks and soil in shades of vibrant green. The plants had the potential to be food, but not without some required changes. The lives of an animal that eats other animals and that of a plant eater are very different, not just in terms of their habits but their physiology. Carnivores have to chase down, catch, and perhaps even dismember their prey before swallowing, but the fats and proteins of another animal body are easy to break down and extract nutrition from. Plants, however, are not so easily digestible. The route to herbivory often requires some significant changes to an organism's body plan, from teeth suited to slicing or grinding vegetation to vat-like guts capable of keeping mashed-up plants inside long enough for symbiotic bacteria to break those plants down. An herbivore's food is virtually everywhere, but that doesn't make it easy to swallow.

Of course, we don't usually find fossil herbivores with their guts intact or even their last meals in their stomachs. Such finds are rare, causing paleontologists to work backward from an herbivore's last meal. Nevertheless, paleontologists can detect certain telltale signs—such as tooth shape—that help identify an animal as more of an herbivore, a carnivore, or an omnivore. And in Carboniferous rocks, one of the oldest apparent herbivores is *Melanedaphodon*.

Roughly speaking, *Melanedaphodon* looks like a weird iguana with a sail on its back. If you're a fossil aficionado, the animal looked like a later, bigger standard of museum displays and toy sets—*Edaphosaurus*. But despite their vaguely reptilian appearance, these early herbivores were more closely related to you and me than to any reptiles. *Melanedaphodon* was an early synapsid, meaning that the back of its skull only had one opening for jaw muscles to pass through. Touch your hand to your cheek bone and you're feeling the outside of that opening in our own skulls. This single-hole configuration is one of the telltale traits that we can follow through the past to detect members of our extended family, meaning those early synapsids were not reptiles but protomammals.

Sadly, we can't watch *Melanedaphodon* crawl around and munch on its preferred foods. But the teeth of the protomammal and its relationship to later plant eaters make it a transitional fossil. The teeth of *Melanedaphodon* were not curved and sharp like that of the famous, carnivorous *Dimetrodon*, but instead were more like blunted pegs. *Melanedaphodon* even had some rounded teeth on its upper palate, predecessors of the broad tooth plates *Edaphosaurus* and related herbivores would later use to mash even greater quantities of plant food. The little protomammal couldn't chew, but instead likely mashed up low-fiber plant material much like some modern lizards do. While even earlier landlubbing herbivores might be uncovered, the discovery of *Melanedaphodon* from a fossil site profuse with the remains of other Carboniferous animals allowed me to focus in on a particular time and place in the period instead of a more diffuse collage.

Appendix IV

I couldn't write a book about paleobotany without some kind of homage to the Chinle Formation, the ancient Mesozoic rocks that host the rainbow shades of Arizona's Petrified Forest National Park and are exposed elsewhere around the Four Corners region of the Southwest. The mineralized conifer trunks scattered among the dazzling desert hues are a natural wonder, and I wanted to use them to speak not only to the ancient lives of plants but also the afterlives of the trees we often find as mineralized stumps.

The American Southwest's Chinle Formation represents a time near the end of the Triassic Period, right around 220 million years ago. The floodplains hosted huge amphibians with heads like toilet seats, gharial-like phytosaurs, an entire group of variously sharp-toothed, armor-plated, and beaked crocodile relatives, not to mention early dinosaurs and their close relatives. Groves of towering conifer trees provided shade and cover to them all, animals and plants found in such close proximity to one another that we can at least begin to outline what this part of Pangaea must have been like during the first act of the Mesozoic.

For this chapter, I chose to combine what paleontologists have learned from not only Petrified Forest but another Chinle locality the next state over in New Mexico called Ghost Ranch. Among the gray rocks just across from the property's main gates are the churned bones of the various creatures who lived in the conifer-dominated forests of the Late Triassic. I've scraped a few dozen bones out of the quarry during field expeditions myself, and the almost ubiquitous presence of charcoal in the rocks always struck me—despite the annoyance of trying to carefully extract what looked like a piece of bone but turned out to only be an ancient piece of burnt wood. The Ghost Ranch quarries formed when heavy rains inundated landscapes that had been swept by prehistoric wildfires, the dried and destabilized soils easily turning into slurries that carried and collected the remains of various creatures that had perished. The relationship between the Triassic animals and plants, the forest's relationship to fire, and the formation's association with petrified wood led me to create a composite vision that could explain the role of forests in creating unusual fossil logjams and also provide a different way to understand petrified wood.

For a long time, the way petrified wood formed was explained in a very simple fashion. All the porous, water-carrying tissues inside trees make them natural sponges. If you submerge or inundate a tree in mineral-laden water for long enough, then those minerals will precipitate out and either coat or replace the tissues of the tree. (There was a time experts thought you could remove the mineral content of petrified logs to reveal the original plant tissue within, but, sadly, this isn't the case, and it's most common for the minerals to thoroughly replace plant tissues rather than simply mold onto them.) But recent research has indicated that the process is not nearly as simple and is much more variable than had been previously assumed. Hydrogen bonds between silica dissolved in water and the organic compounds of a tree's cell walls is the critical first step that forms a kind of mineral base for additional dissolved silica to build up inside a tree's tissues. The minerals not only attach to the cell walls, but eventually build up inside, too, and over time can create forms of quartz called chalcedony or even opal, the geologic particulars that give a great deal of fossil wood its rainbow colors.

I chose *Silesaurus* as our animal star of this chapter to signal the strange nature of Triassic time, when the earliest members of many familiar groups of animals—including dinosaurs—were just getting their start. *Silesaurus* wasn't quite a dinosaur, instead resembling the sort of animal dinosaurs evolved from even as they continued to persist alongside the terrible lizards of the Triassic. The bones of such reptiles are found at both Petrified Forest National Park and Ghost Ranch's Hayden Quarry, and the omnivorous reptile seemed a good connecting point between the locales and the conditions that led swaths of Triassic forest to become enclosed in the fossil record.

The critical plant in this chapter is *Araucarioxylon arizonicum*, the fossil plant found in such gorgeous abundance in Petrified Forest National Park. Despite the number of large, multicolored logs littering the desert, however, we know relatively little about this tree. The gymnosperm's leaves have yet to be found, and often we see the interior of the tree without the outer bark. Nevertheless, I thought that the fame of *Araucarioxylon* provided an excellent opportunity to dig into how petrified wood is formed. Recent investigations have revealed that the transformation process takes place over a much longer time frame than previously expected and, in the case of the Triassic trees, is reliant upon dissolved silica in the water. Silica is very common in the Earth's crust, and therefore Earth's water, and as water doused buried trees, silica began to build up on the tissues. Once an initial rime of silica formed, it became easier for more silica to infiltrate the tree tissues and preserve it. The rainbow colors of petrified wood depend on other minerals and chemical compounds in the groundwater the tree was exposed to. The varying hues of petrified wood through time are an indicator of what was dissolved in the ancient groundwater, a geochemical look at ecosystems we can't visit directly.

Appendix V

The Morrison Formation is dinosaur country. Stretching from southern Canada nearly to the Mexican border, as wide as the span between Utah and Oklahoma, this 10-million-year stack of geologic time contains the remnants of one of the most unusual and fascinating ecosystems to have ever evolved. These rocks not only contain the bones and tracks of famed dinosaurs such as *Apatosaurus*, *Allosaurus*, *Stegosaurus*, *Diplodocus*, *Supersaurus*, and more, but the habitats that allowed these amazing saurians and their neighbors to thrive. I visit almost every summer, hoping to find new fossil sites around the historic Cleveland-Lloyd Dinosaur Quarry in central Utah. To understand these interactions, however, paleontologists often have to work backward.

Even though the Morrison Formation is world-famous for its dinosaurs, plant fossils are harder to come by. Most are the petrified trunks of conifers, with leaves being relatively rare. Chance discoveries, such as a Jurassic "Salad Bar" locality found outside Dinosaur National Monument, are slowly filling in what we understand about what plant species lived in these habitats. Sometimes even a partial leaf or piece of trunk is enough to let paleobotanists know what sorts of plants were around. Piecemeal, paleobotanists have come to understand that the Morrison Formation hosted a diversity of ferns, ginkgoes, conifers, shrub-like podocarps, and horsetails, among others. There were no grasses or flowering plants here. Those would not even begin to evolve for about another 25 million years after the time of the Morrison Formation. Instead, the landscape was clothed in much more ancient forms of plants that had been diversifying alongside the dinosaurs for tens of millions of years.

Paleobotanists used to think that these plants were not very nutritious. That left something of a puzzle. How could Morrison Formation dinosaurs like *Supersaurus*—which could get to be over one hundred feet long and weigh more than forty-five tons—get so large on such low-quality forage? It turns out that part of the answer involved checking our assumptions. Even though botanists had long assumed that flowering plants offer much more energy and nutrition to animals than plants like ferns and monkey puzzle trees, laboratory studies have changed that view. Many Morrison Formation plants have living relatives with similar anatomy. By "digesting" these plants in laboratory settings, paleobotanists have found that ferns, horsetails, monkey puzzle trees, and especially ginkgoes would have provided adequate food for dinosaurs like adult *Apatosaurus*.

Naturally, part of this evolutionary dance involves the dinosaurs themselves. We know from sauropod skulls of different ages—such as juvenile and adult *Diplodocus*—that young sauropods had rounded snouts best suited to selectively browsing their greens, picking out the most energy-packed foods to fuel their incredible growth spurts. These dinosaurs didn't take a century to become giants. They had to grow fast in order to protect themselves from the large carnivores that lived in the Morrison Formation, part of an evolutionary back-and-forth told through the patterns of growth detected inside dinosaur bones. But adult sauropods have broad, squared-off muzzles, which is typical of grazers. We know from studies of living animals that there are trade-offs between body size and diet. A very large animal—like an African elephant or an *Apatosaurus*—has to eat more in absolute terms but can get by on lesser-quality food than a much smaller animal that can eat less in absolute mass but needs to find fruits, seeds, or other high-energy foods to survive. What this all comes down to is an evolutionary dance between herbivores and plants, with the available forage opening the possibility for dinosaurs to become so impressively huge. Without the right kind of plant food, the sauropods would have never gotten so big. The fact that Jurassic plants such as horsetails and ginkgoes have biological insurance policies—such as underground root networks and stashes of starch among their nascent buds—indicates that these plants evolved to respond with new growth when the green parts were eaten, distressed, or otherwise disturbed, both allowing their survival and providing plenty of fodder for the dinosaurs.

Naturally, a tree like the *Ginkgoites* in the latter half of the story aren't just isolated organisms. They are ecosystems in and of themselves, hosting and feeding various other species. Fossil plant leaves from the "Salad Bar" site have insect damage, perhaps caused by Morrison Formation beetles or caterpillars. Likewise, anyone who has been around a living *Ginkgo biloba* tree when it's fruiting knows that they *stink*. That's not to drive us away. It's to attract animals like tanuki that want to ingest the fruits because they smell like rotting meat, the animals then carrying the seeds away from the parent tree in their digestive systems and depositing them in their scat elsewhere. Multituberculate mammals are often reconstructed as very squirrel like, and several have been found in the Morrison Formation—*Ctenacodon* being one of the best-known of these little mammals. While we can't know what a Jurassic *Ginkgoites* smelled like, it's a fair supposition that these trees relied on animals—like dinosaurs or multituberculates—to swallow and spread the tree seeds, and so *Ctenacodon* plays the role in our tale that raccoon dogs do today.

Appendix VI

In 1879, Charles Darwin coined a phrase that would hang over the heads of botanists ever after. Thinking through time, the English naturalist was puzzled by the seemingly quick evolution of flowering plants—or what are technically called angiosperms. Still locked into thinking of life-forms as "lower" or "higher" based upon their perceived complexity, Darwin wrote that "the rapid development as far as we can judge of all the higher plants within recent geological times is an abominable mystery."

Darwin didn't really have a full concept of Deep Time when he was writing. He couldn't have. Radiometric dating and investigations into the absolute age of the Earth were decades away. All geologists could work from were relative ages as displayed in rock strata, with newer life-forms appearing in higher rock layers and older ones in those below. Nevertheless, the big picture was that flowering plants were latecomers in comparison to conifers, ginkgoes, ferns, and other plants, and it seemed that angiosperms would have needed some kind of edge—like help from pollinators—to spread so fast.

Now we know that Darwin's abominable mystery wasn't really an unexpected event that requires special explanation. Insects became plant pollinators millions of years before angiosperms or attractive flowers became prominent, for one thing, and angiosperms are much older than Victorian scientists could have perceived.

Identifying the earliest angiosperm is a fraught task for a few reasons. The first is that evolution presents us with a rough continuum, with new species spinning off from those already established. If we had a totally complete fossil record it would be hard to tell where one species stops and the next one starts. But even when embracing the imperfection of the fossil record, and everything we have yet to find, the fact is that many plant fossils require particular circumstances to become preserved—especially less-resilient parts of the plant like the leaves and flowers. No matter what the current contender for oldest angiosperm is, we know that there has to be something even older that we either haven't found or just did not become preserved.

Paleobotanists have been debating the identity of the true oldest angiosperms for a long time. Some molecular studies have proposed that the earliest angiosperms must have evolved by 320 million years ago, in the age of the scale trees. There are at least a handful of papers on angiosperm-like pollen from the Triassic, the beginning of the Age of Reptiles.

Early Jurassic plants found in Nanjing, China, have been proposed as the earliest known flowers, too, only to turn out to be conifer cones. It's not impossible that angiosperms existed earlier than we presently know, but so far the fossil evidence leads us to the Yixian Formation of China. The angiosperms found there were not the first to have evolved, but they remain among the oldest yet known.

Treating the term angiosperm as a synonym for "flowering plant" is a bit misleading, looking at the modern plants we most closely associate with flowers and following that thread back to its origin. The truth is that other, non-angiosperm plants produce fruit and flower-like structures of their own, and not all angiosperms flower. It's even caused some professional confusion with some flower-like fossil plants being mistaken for early angiosperms even though the details of their anatomy mark them as members of other groups. Reconstructed as it would have looked in life, *Archaefructus* looked like a wispy aquarium plant more than most plants we currently associate with flowers. And while *Archaefructus* was originally described as a kind of proto-angiosperm, recent research has found that it was a true angiosperm that was already becoming somewhat specialized. I used the fact that it's found in water, while the vast majority of living gymnosperms are hydrophobic and don't grow in wet soils, to suggest that calm ponds and lakes might have provided an opportunity for the relative newcomers to find places to grow where they wouldn't have to compete so strongly with the existing conifers and cycads.

The notion that *Archaefructus* represents one form of early angiosperm rather than what the ancestor of angiosperms looked like comes from another source that was published while I was in the process of writing this book. It's always a wonderful feeling when something directly relevant to your story is announced right in the middle of it, even if it requires some revisions. A fossil of the fossil bird *Jeholornis* has preserved phytoliths in its guts, or mineralized plant parts that are often resilient to digestion. The phytoliths resemble those of trees related to modern magnolias, indicating that there were terrestrial angiosperm trees around about 125 million years ago, too. Given that *Jeholornis* is found in the same geologic formation as *Archaefructus*, including the plant-eating bird felt like a good way to expand on the early spread of angiosperms even as paleobotanists continue to wonder when, where, and how the first of these plants evolved.

The rocks of the Yixian Formation preserve much more than plants and birds. The ancient layers are brimming with delicately preserved fossils, including beautiful birds, pterosaurs, non-avian dinosaurs, insects, and other forms of life in addition to the flora. The hadrosaur *Jinzhousaurus* comes from these rock layers and is a good candidate for highlighting a long-running debate in paleontology.

The fact that angiosperms arose in the middle of the Age of Dinosaurs piqued paleontologists' curiosity. The spread of new plants that produce an abundance of fruit and flowers raised the question of whether dinosaurs might have "invented flowers." In the Northern Hemisphere, in particular, landscapes that once hosted a variety of long-necked sauropod dinosaurs that scarfed plants without chewing were often replaced with dinosaurs like hadrosaurs and horned dinosaurs capable of actually chewing plants. Perhaps dinosaurs being able to better process plants and the comparatively fast growth rates of angiosperms were connected, the dinosaurs

inadvertently giving flowering plants an edge. But, roman-
tic as it is, dinosaurs did not invent flowers. The timing isn't
right, and the respective changes among both dinosaurs and
plants stem from different sources. Angiosperms did not immediately take over the landscape, and pollinators like beetles
and birds had more to do with their evolutionary flowering
than dinosaurs. And among dinosaurs, early hadrosaurs like
Jinzhousaurus represent lineages that were once on the ecological sidelines undergoing a radiation in landscapes where
sauropods had been decimated or disappeared. Dinosaurs capable of chewing were better equipped to break down fibrous
and tough vegetation, including rotting logs where hiding
snails and crustaceans would have provided the munching
dinosaurs with some extra protein. What *Jinzhousaurus* represents is a shift in the ways dinosaurs interacted with the
landscape, making meals out of plants that were likely ignored
by giants that had to eat as many soft, calorie-rich plants as
they could to fuel their stupendous growth.

The idea that *Jinzhousaurus* also munched on rotting logs
comes from later fossils found in different places. Fossilized
dung from other ornithopod dinosaurs, likely duck-billed dinosaurs, found in southern Utah, show that the dinosaurs
used their impressive teeth and jaws to munch on rotting
logs. The petrified plops contain rotted wood fibers as well as
snails and small crustaceans. We don't have direct evidence
that *Jinzhousaurus* did the same, but there's also no reason to
think that the herbivore wouldn't have made the most of decaying tree trunks and the extra protein found within. Likewise, the vignette about the deceased sauropod comes from
fieldwork I've carried out over the past decade. Often, fossil
dinosaur bones near the surface are damaged by the roots of

juniper, sage, and other plants that leach minerals out of the fossil bones. Paleontologists often have to cut roots and repair bones that have been busted up by roots growing into them. The same may have been true in the Mesozoic, the bones of dinosaurs buried near the surface providing an important mineral source for plants. I included the speculation to stress the importance of how life and death have a complex relationship in ecosystems both ancient and modern, and even the afterlife of a living thing can help nourish what survives them.

Appendix VII

Sometimes I wonder what some paleontologist of the future might find locked in amber from our present moment in time. A tuft of deer fur, a mayfly that perished too early in its all-too-brief life, a mosquito worthy of a *Jurassic Park* close-up, all these things and more are possible. Right now, somewhere in a wood, a piece of our world is being encased in preternaturally sticky tree ooze, a time capsule that will likely last longer than our species.

Amber, simply put, is hardened tree resin. It's not just sap. All trees create some kind of sticky sap inside their branches and trunks, but the fluid is thin and a little watery. You might have noticed some on your hands while weeding or fussing in the garden, an equivalent of plant blood that helps carry around hormones, sugars, and other essential ingredients to a plant's life. But resin is something special. Resin is thicker, darker, and stickier—if you've ever gotten some on your clothes, you know—and acts as a natural bandage. The role of resin isn't to transport nutrients and water around a plant. It's to gum up a wound in the bark, filling in the gash so that insects, fungus, or other organisms that might harm the tree can't get in. And among living trees, only *Pinaceae* trees—think cedars, firs, pines—create resin. All amber gets its start as extraordinarily gloppy resin oozing from an injured tree.

Paleobotanists aren't certain if only prehistoric pines created resin or if other, now-extinct trees also suppurated the tacky stuff. Just like paleontologists usually don't find the body of the track-maker with a fossil footprint, amber is usually out of context from the tree that made it. Nevertheless, amber is a blessing for anyone curious about the past. Primordial trees that oozed resin collected the remains of living things in their environment, offering us details and evidence of interactions that we would otherwise entirely miss. A tiny ammonite shell—the home of a cephalopod that lived in the ocean—encased in amber must have started as resin produced by a tree along a Cretaceous coastline, for example, and paleontologists have found evidence that prehistoric dermestid beetles once lived in dinosaur nests thanks to chewed-up dinosaur feathers and beetle larvae casings preserved in ancient resin.

Contrary to what you might expect from a certain blockbuster movie franchise, amber is not like sticking a prehistoric creature in a vacuum seal bag and unsealing them fresh as the day they were preserved millions of years later. Prehistoric organisms in amber are often degraded and distorted, smooshed by all the time in the rock and never 100 percent complete. Even a little bird or insect that might look pristine and complete from the outside is usually missing their internal organs and other biological details. And the same goes for prehistoric DNA. I am entirely certain that I will spend the rest of my fossiliferous career answering this question, but, sadly no, there is no hope of recovering prehistoric DNA from amber-encased insects or other fossils. The problem isn't technological. Paleontologists have developed techniques to extract ancient DNA from bones, teeth, and even the prehistoric soil that ancient species walked over. No, the problem is that DNA has a kind of half-life. Genetic material begins degrading at death, no matter how nestled a living thing might be by amber, and even under ideal conditions all that biological unspooling occurs fast enough that any original DNA will be broken down within about 6 million years. We might be able to get tattered DNA from a Pliocene sabercat that fell in a cave, or even a very early human, but non-avian dinosaurs and other organisms in millions-of-years-old amber are pretty much out of the question. Whatever we're going to learn about the DNA of *T. rex* and kin, we're going to have to figure out from fossil genes living in modern birds.

Of course, even if *Jurassic Park* were possible, the menagerie of dinosaurs and other living things would be fundamentally different from any collection of Mesozoic reptiles we currently know about. We wouldn't have *Tyrannosaurus*, *Brachiosaurus*, and *Velociraptor*. We'd likely have entirely new, unheard of dinosaurs, perhaps even from groups we haven't found yet. Prehistoric amber deposits are found in diverse places around the world of varying ages, often in geologic formations that lack body fossils. In other words, amber formed in environments that did not favor the preservation of skeletons—especially not of big animals like our favorite dinosaurs.

The oldest amber paleontologists know about comes from the Carboniferous, and there are more recent amber deposits from the Cenozoic, too. But there was never a better time for amber formation than between about 125 and 70 million years ago, a time that an international team of paleontologists has dubbed the Cretaceous Resinous Interval. A combination of different factors ranging from the presence of insects capable of damaging trees to volcanic activity that warmed the global climate all played into this peculiar time, a span of Cretaceous time when amber was being formed in much greater quantities than ever before. The key was a warm, humid climate. Resin-producing conifers and other plants were able to grow over much broader ranges, including in environments like floodplains where sediment was being laid down. In these lowlands, a greater amount of resin resulted in greater chances of a bioinclusion—an organism or its parts becoming trapped in resin—and that piece of hardened resin being buried. Most of the time, paleontologists have found, amber with bioinclusions end up being transported by water—perhaps by seasonal rains that send water running over the landscape and picked up the hardened resin like pebbles—but sometimes they find it where the resin was initially formed.

It's unlikely that a similar resinous interval will occur again. Angiosperms are much more widespread and diverse than gymnosperms are today, even though some angiosperms create amber-producing resins, too. The conifers responsible for so much of the world's amber were able to do well in a warm Cretaceous world where angiosperms were just beginning to become more important parts of the landscape, a condition that would require angiosperms to go through a mass extinction of their own to allow those resinous trees a chance to grow again. Amber with bioinclusions primarily come from Cretaceous rocks, making them rare even compared to other fossils. And this is where the ethical issue starts to come into play.

We love amber. People have been collecting and making jewelry out of amber for centuries, and amber pieces with insects or other fossils inside are the most prized of all. Over time, the amber market has turned into a highly commercial enterprise that largely operates away from public view. That's partly by design. In recent years paleontologists and amber collectors alike have been dazzled by amber fossils coming from the roughly 99-million-year-old rock of Myanmar. But that amber comes at too high a price.

Myanmar's amber is so highly prized that the military government seized the country's amber mines in 2019, many of the fossils from the site making their way to China where they are sold on the black market. The cash goes back to fund the genocidal conflict in Myanmar. The fossils have become known as "blood amber," and groups such as the Society of Vertebrate Paleontology have pushed for a ban on all research and publication on Myanmar amber collected from 2017 onward, as well as greater transparency into how those amber pieces are obtained and placed in museums. Too often have amber fossils purchased without any background information been acquired by private collectors, placed in museums that cannot properly care for the amber, and then used to fuel academic careers while the people of Myanmar suffer. All fossils come from somewhere, after all, and prehistoric creatures being sold as trinkets for the rich is not some harmless eccentricity.

I could have selected amber from another locality from the 55-million-year interval when more resin flowed than any time before or since. But doing so felt like sidestepping the ethical considerations that paleontology is only just now beginning to grapple with. Paleontology has benefitted greatly from colonialism and a view that fossils are primarily scientific objects, ignoring their connection to modern places and peoples. The market for blood amber has not only funded deadly conflict, but also exported Myanmar's natural history heritage with untold numbers of specimens sold off to private collectors looking for a unique curio to remind them of their wealth. By exploring the forests that oozed Myanmar's amber, I hoped to call additional attention to the ethical complexities of a nineteenth-century science meeting twenty-first–century realities.

Appendix VIII

When we talk about life after the asteroid impact that brought the Cretaceous to a catastrophic close, we often focus on animal life. The dinosaurs were lost, save for beaked birds, the flying pterosaurs eradicated, mosasaurs never to be seen again, and even the extremely prolific ammonites faltered about 100,000 years after the disaster. Not only do we take the story of life's recovery as a given, but plants are usually ignored—just the backdrop. After all, how much difference could there be between a forest of 66 million years ago and its replacement of 65 million years ago?

But there was a mass extinction of plants just as there was of animals at the end of the Cretaceous. Even though flowering plants had been around for millions of years prior to impact, and the world even saw grasses like palms start to become more common, the vegetal majority on Late Cretaceous Earth was ferns, gymnosperms, and other plants that formed the foundation of terrestrial ecology since the Triassic. There were no modern-looking jungles or forests in the time of

Tyrannosaurus and *Triceratops*. It was only after the extinction's pressure eased that modern-looking forests began to grow, including the first examples of what we call neotropical rainforests.

I briefly described what the forests of early Paleocene Montana might have looked like in the final narrative chapter of my previous book, *The Last Days of the Dinosaurs*. Its characterization was largely informed by discoveries at a site called Corral Bluffs near Denver, Colorado, perhaps our best record yet of the million years after impact. I was tempted to return to those fossils and write their story anew, with greater focus on the forest, but I decided to pick a different example for this pivotal moment in life's history. Fortunately for me, recent discoveries in the Paleocene rocks of Colombia have provided a glimpse into how forests underwrote life's comeback after the extinction that ended the Cretaceous.

Unsurprisingly, megafauna tend to get more attention than botanical finds. In 2009 paleontologists announced the discovery of the largest fossil snake yet known, *Titanoboa cerrejonensis*, from a coal mine in La Guajira, Colombia. A constrictor that grew more than forty feet in length—a suitable length for a reptilian movie monster—certainly grabbed the world's attention. But the setting in which *Titanoboa* was found had a great deal more to tell us about the Paleocene world than the snake. The coal of the mine is the remnants of a tropical forest that *Titanoboa* slithered within 60 million years ago, a jungle vastly different from the conifer-rich woodlands that the great dinosaurs knew.

The coal mine where *Titanoboa* was found sits within the Cerrejón Formation. Pollen and plant leaves alike, representing more than a thousand species of plants between the Cretaceous and Paleocene, indicate that it took about 6 million years for plant communities to build up to pre-extinction diversity.

The fact we know anything at all about the origins and deep history of neotropical forests is a wonder. The soils of modern rainforests often trend toward the acidic. When plants are growing enthusiastically, even their roots can increase the amount of carbon dioxide in the soil. As rain showers these forests and keeps them wet, chemical reactions between the rainwater and soil composition cause the water to pick up some of that carbon dioxide and become more acidic. The interaction has major implications for what grows in rainforests—and why, despite their sprays of greenery, rainforest soils are generally bad for growing crops—but also for fossilization. Acidic soils are unkind to everything from bones to leaves buried by the local sediments, requiring particular conditions to be wrapped up in the fossil record before being entirely destroyed. So to find not only the body fossils of animals like *Titanoboa*—but also thousands of plant fossils—in the Cerrejón Formation is a lucky break for paleontologists.

I also wanted to take the opportunity to include some new research on how plants grow toward light. Botanists have long known that plants grow and bend toward the light. The mechanism for how plants are able to do this has been a mystery. While composing this book, however, botanists published a new study that found at least one way plants can detect where the sun is. In a laboratory experiment, almost all seedlings in a sample grew toward the light as expected. One didn't. When the researchers investigated the unusual plant, they found that what should have been air spaces in the plant were filled with water due to a mutation. The plant could not seem to find the sun and didn't grow in a consistent direction.

Through further analysis, the botanists realized that light changed direction as it passed through the growing plants. When the light passed from the water-rich cells into the air spaces, it refracted as happens any time light moves from one medium to another. The mutant plant was so waterlogged that the light didn't scatter and therefore was not giving the plant crucial information. So far, the research has only been carried out in seedlings and may not apply to all plants. The finding is so important that I felt it belonged in the book, however, and so imagining how swampy palms might track the sun seemed a good opportunity to incorporate the hypothesis.

Appendix IX

South and Central America are home to dozens of species of unique primates, different from any others in the world. Primatologists know them as platyrrhines, the capuchins, spider monkeys, and marmosets that form their own evolutionary group separate from their primate relatives in Africa and Asia. That split has a very ancient origin. Sometime before 32 million years ago, multiple lineages of monkeys must have made the journey between Africa and South America. The best candidate for their voyage? Rafting on vegetation torn up from the ancient African coast.

For decades, the idea that monkeys somehow traveled across the ocean to South America was based on conjecture. Even though there were early primates present in North America at the start of the Cenozoic, those archaic lineages went extinct in the ancient Americas. Nor was there any indication that a relict population somehow survived in South

America while the continent was an island—the continent was cut off from other landmasses by the end of the Eocene, not to be reconnected until Earth's tectonic plates brought it into contact with what's now Panama about 9 million years ago. The ancestors of today's howler monkeys had to have arrived sometime within that window, and by the 1970s the leading hypothesis was that they must have rafted there.

The monkeys weren't the only clue that something unusual must have happened. The fossil relatives of South America's guinea pigs and capybaras, called caviomorphs, resembled those found in Africa even though there was no overland route for them to travel between the landmasses. There were no land bridges, and any route via other continents would have taken millions of years, leaving a fossil trail behind. And these were not isolated cases. Madagascar had split from mainland Africa by the time lemurs and tenrecs arrived there, rafting being the only sensible way for the mammals to have crossed the channel between the island and the continent. And animals manage to island-hop by rafting today, too. Hurricanes in the Bahamas have washed anole lizards into the sea, sometimes on vegetation and sometimes simply afloat in the saltwater, and enough lizards have survived these journeys that biologists can detect where they came from and how they affected the genetics of island populations through their DNA. Despite the reputation of islands to be evolutionary "laboratories," insulated spaces where adaptation can unfold without input from the outside world, the truth is that channels and even oceans are not the barriers they were previously thought to be.

It wasn't until 2015, however, that paleontologists began to find evidence of South America's first primates. The fossils hinted that what had transpired was more complex than anyone had previously thought. The 36-million-year-old *Perupithecus*, which extended the fossil record of the continent's monkeys back by 10 million years, didn't show any direct resemblance to modern platyrrhines. The announcement of *Ucayalipithecus* in 2020 indicated that a totally extinct African primate group called parapithecids had arrived in South America around this time, as well, meaning that multiple ocean crossings was likely. Then, in 2023, paleontologists found the tooth of another early South American primate, *Ashaninkacebus*, that could either belong to an early primate group called eosimiids or be related to the ancestry of platyrrhines. At least two, if not three, primate groups had crossed the ancient Atlantic, all at some time prior to 36 million years ago.

Experts had focused for decades on platyrrhine monkeys and their ancestors. No one had considered that multiple lineages of African primates might have made the same journey. What was supposed to be a one-in-a-million chance might have actually occurred multiple times, perhaps in a window where shifts in sea levels or an uptick in intense tropical storms repeated the process enough times to turn the journey into more of a numbers game than a fluke.

It'd be foolish to say direct evidence of such crossings will never be found. Oceans and coasts harbor many places where new sediment settles. It's entirely possible that an unfortunate raft of primates or rodents sank and became preserved in ocean rocks from somewhere along the route between the continents. There's no way to look for it directly, worse odds than finding a needle in a haystack, but it's not beyond the realm of possibility. All the same, the narrative of this chapter is formed around what we know from the ecology of prehistoric South America and Africa, as well as clues from modern primates.

While there may be more complete fossils of *Perupithecus* and its early primate neighbors, their fossils are often found on riverbanks that are very difficult—even dangerous—to work. Paleontologists often scoop shovelfuls of soil and then screen-wash the loose sediment away, revealing teeth or whatever hard parts survived the process. It's a kind of compromise, a method that allows for new discoveries but also likely destroys small bones that are not as resilient as fossil teeth. I based my characterization of the primates in this chapter on common behaviors among modern primates, admittedly projecting the present onto the past, but a necessary step given the story would otherwise be about the travels of teeth between continents.

Reconstructing what the crossings between prehistoric Africa and South America must have been like is a challenging task. The primates saw and experienced things we will never know. But we can assume that the journey must have been relatively calm, unhindered by storms that might break up or submerge the floating rafts of vegetation. I also had to wonder if any oceangoing creatures would have visited the rafts, curious of what they were. Marine mammals were beginning to thrive around 40 million years ago, including early whales and the ancestors of today's manatees and dugongs. I decided to use a spout as a way to raise the prospect of a menacing surprise guest—the huge predatory whale *Basilosaurus*—before revealing the mammal to be an early, limbed manatee. For this portion, I took my inspiration from *Prorastomus*, a manatee relative from time-equivalent rocks in what is now Jamaica. We still don't know very much about it beyond the skull, a few neck vertebrae, and additional parts. Based upon the time it lived, and the more complete remains of another sirenian called *Pezosiren*, it's likely that these beasts still had discernible arms and legs around 40 million years ago and were capable of walking on land as well as swimming. But the fact that *Prorastomus* was found in Jamaica, across the sea from some of its closest relatives in Eocene Africa, indicates that sirenians were already capable of swimming across entire oceans long before their tail fins evolved. Whales show much the same pattern, dispersing to distant coasts while they still swam with their limbs. The distance between prehistoric South America and Jamaica led me to avoid calling the manatee in the chapter *Prorastomus*, as it's likely a different species lived along those coasts, but the fact that sirenians had already crossed the Atlantic hints that there would have been sea cows munching on sea grass along prehistoric South America, too.

Both in the water and on land, South America was not as isolated as traditionally depicted in paleo media. The twentieth-century paleontologist George Gaylord Simpson wrote an entire book called *Splendid Isolation* about the ways fossil animals diverged from their North American counterparts during the time South America was an island, between 130 and 4 million years ago. And it's true that a land of terrestrial interchange shaped evolution on the landmass, especially because the joining of the Americas meant that giant sloths and armadillos moved northward while cats, dogs, camels, elephants, and other creatures ventured south in the Great Biotic Interchange. Nevertheless, no landmass is ever entirely isolated. We've long known that seeds of plants like screwpines can float to new land across the seas, that birds carry small seeds and other hitchhikers with them, and that chance events—like rafts of primates—can introduce new animals to distant places. Improbable as it may sound, accidents have played essential roles in where living things thrive.

Appendix X

It's difficult to overstate the importance of grass. It's a group of plants that seems encompassing today, perhaps as close as the front yard or the tuffet you're sitting on if you're reading this book outside, but the ubiquity of the green stalks is something relatively new. The spread of grasses has had huge consequences for the history of life, altering the evolution of herbivores and the carnivores that feed upon them.

The oldest fossil grasses we know of are about 100 million years old. They grew in the cracks of the Cretaceous world along with closely related fibrous plants like palm trees. But there were no grasslands during the heyday of the non-avian dinosaurs. *Triceratops* didn't herd like bison across the plains. It wouldn't be until about 20 million years ago or so that global climate shifts favored a drier global climate that ate away at the thick semitropical forests that sprung up in the Cenozoic's hothouse years, increasingly spread out by patches of grass and low-growing plants.

I decided to go back a little bit further to set up this critical moment in Earth's botanical history, to the end of the Eocene around 34 million years ago. Rather than bring you into a time when single-toed horses were running fast over hard-packed grassy plains, I wanted to focus on an ecological turning point where the effects on herbivorous mammals could be better seen. The relationship between brontotheres and horses seemed a perfect way to juxtapose the way plants shape the histories of herbivores—and all the lives they are connected to.

Museums love brontotheres. Many of the "thunder beasts" are large, impressive, and rhino-like without actually being rhinos, adding to the sense of an exotic past. The big species, like *Megacerops* in this chapter, are the most famous. But paleontologists have recently found that brontotheres thrived at both large and small sizes during their time, comparatively giant and short-statured brontotheres evolving alongside each other among prehistoric forests. They were beasts most at home within the humid, thick forests of Earth's warm period, and their skulls underscore the point. Animals that browse on soft leaves and less-fibrous plants tend to have low-crowned teeth and rounded muzzles, the hallmarks of browsers that clip leaves. Grazers, by contrast, tend to have high-crowned teeth and squared-off muzzles to bulk-feed on low-growing, abundant vegetation—the kind of profile seen in dinosaurs like *Diplodocus* in the Jurassic and cows today.

Brontotheres happened to be part of a major mammal group that did very well in the first half of the Cenozoic. Perissodactyls are "odd-toed ungulates," represented by horses and rhinos and tapirs today, that stand on one or three toes. Their center of balance is usually a line straight down their middle toe on each foot, allowing the side toes to spread out and support the animal's body weight. It's a great arrangement for habitats with muddy, mucky soils that stay relatively wet. Artiodactyls, or mammals that stand on an even number of toes and eventually evolved dainty, deer-like feet, were not as common during these times. You were more likely to see a brontothere, a rhino, a tapir, or a chalicothere—imagine a horse with multiple clawed toes—than a relative of antelope or camels. Horses were among this perissodactyl profusion, although they didn't look very much like the stallions and mares we know today.

The earliest horses were famously small and had multiple toes. *Eohippus*, the "dawn horse" of 55 million years ago, is the classic example. They were inhabitants of the forest, much like the brontotheres were. Even as horses became larger and stood on three toes per foot, their bodies were best suited to turning forest leaves into fuel. But by the end of the Eocene, horses were beginning to change. *Mesohippus* still had three toes per foot and low-crowned teeth, but the herbivore's body weight was balanced along the central toe—the equivalent of our middle fingers and toes—and indicates a change. Brontotheres remained tied to the forests while small horses began to forage among the more open patches of habitat, and this is where the carnivores come into the picture.

Carnivores that live in forests are often ambush predators. Think cats and their kin. These meat-eaters rely on stealth and getting close to prey before a pounce and a bite. More open landscapes led to the evolution of pursuit predators, more dog-like forms whose limbs were less suited to grappling and more toward the forward-back motion for efficient running. If you have a dog and a cat at home, think about the difference in how they move and play. Your cat is an ambush hunter while your dog loves walks and runs because their ancestors were pursuit predators. And back in the Eocene, as horses were emerging from the edge of the forest, the carnivores waiting there nudged the herbivores toward what we think of today when we hear the word "horse." Rather than becoming larger to deter meat-eaters, as some brontotheres and rhinos did, *Mesohippus* and other horses became quick, shifting their anatomy to make them better runners. You can see it in their limbs. Animals that run fast usually have very short upper parts of the leg but very long lower leg and toe bones that create more reach. Early horses with longer lower legs had a speed advantage, allowing them to survive longer to pass on those variations, and over time those side toes stopped touching the ground. If anything, they began to become a liability— little splints that might be broken in a fall or a wrong step. The process took millions upon millions of years to play out, but we can see the beginnings of these shifts in *Mesohippus*.

Grasses also required new teeth. It's a passive defense, like armor, but many grasses are fibrous and packed with silica. Not to mention an herbivore pulling grass out of the ground is going to eat a great deal more sediment than one munching on leaves held over the soil. It's a big problem. Dental damage can be deadly. An animal that wears down their teeth too fast might become incapable of feeding while relatively young, and eroded teeth can expose parts of the tooth to bacterial infection that can kill. Grass, like predators, presented an evolutionary push, and so after the Eocene some of Earth's herbivorous mammals started evolving different tooth arrangements. Early horses, for example, still kept the standard arrangement of one set of milk teeth replaced by one set of adult teeth, but those teeth became much longer. If you look at an X-ray of a modern horse's jaw—domesticated or wild— you'll see that their tooth crowns extend far down into the jaw itself. As they grind down their molars, the crowns move up, almost like a push-up popsicle. These changes are a direct consequence of becoming suited to grasslands, and so the juxtaposition of one of the last brontotheres with a horse on the cusp of major changes seemed a fitting way to contrast how changes in vegetation rippled outward.

Appendix XI

Pollination is a very close, even intimate, interaction between species. Animals of one form or another have been helping plants to spread their gametes around their world since the heyday of the dinosaurs, even before the origin of angiosperms or brightly colored flowers.

So far, the earliest candidates for plant pollinators go all the way back over 201 million years ago to the Triassic. Paleontologists working in Russia found ancient insects called tillyardembiids—ancient earwig relatives—that were absolutely doused in prehistoric pollen. The little granules couldn't have come from Triassic flowers because anything like the blooms we see in the garden hadn't evolved yet. The pollen must have come from plants around at the time, and the details of the pollen covering the insects suggests that it primarily came from conifer relatives called cordaitales with strap-like leaves. Insects so thoroughly draped in pollen must have been capable of transferring pollen between plants, inadvertently helping the plants fertilize each other instead of relying on the wind alone. Early pollination likely looked much like this, with unknowing insects carrying pollen between plants as the bugs clambered over them.

The tillyardembiids aren't the only prehistoric animals likely to be pollinators. Paleontologists have identified pollinators from various insect groups, including some—like scorpionflies—that don't have pollinators among their living species. Some, such as a beetle encased in 99-million-year-old amber, are covered in pollen while others are suspected on the basis of mouthparts of body types associated with pollination. But we also know birds and mammals play important roles as pollinators in our world today, especially bats. Here, the fossil record gets a little trickier.

Most of what we know about fossil birds and bats comes from their skeletons, and often only portions of those skeletons. Small flying animals with delicate bones don't preserve especially well in the fossil record unless there are truly superb circumstances. Still, a fossil bird from the 47-million-year-old rock of Germany was found with pollen grains from an angiosperm in its gut contents—a strong hint that this bird visited flowers and helped pollinate them. And while such direct evidence has yet to be found for fossil bats, it's highly likely that the flying mammals have been doing so for a very long time.

I based the primary vignette in this chapter on a small bat uncovered on New Zealand's South Island. The little fluttering mammal lived there between 19 and 16 million years ago, yet is the oldest representative of a bat genus that's still around today—*Mystacina*. The bats aren't quite like the stereotype of little brown mammals roosting in caves and swooping through the night. *Mystacina* bats often forage on the ground—shambling after insects on folded wings—as well as flying through the air. And when flowers are in bloom, *Mystacina* are important pollinators.

While it's unlikely paleontologists will find a fossil of a pollen-covered Miocene *Mystacina*, the paleontologists who described the ancient bat found that relatives of the plants the bats pollinate today were also growing in the same forests where prehistoric *Mystacina* lived. It's possible that the relationship between the living bats and the plants they pollinate goes back a long, long way, an echo of life as it was when sabercats and shovel-tusked elephants roamed other parts of the world.

The existence of a liana that lost its pollinator is conjecture based upon the fate of some modern relatives in New Zealand. The kiekie is a modern liana that is thought to have been primarily pollinated by bats but still hangs on in places where the specific bats have disappeared. Botanists have determined that the bats were not exclusive kiekie pollinators, with birds chipping in, but the plant does seem to struggle with dispersing its seeds in places where the proper bats aren't present. Instead, possums that have been introduced to some parts of New Zealand have accidentally taken over as pollinators. They eat the flowers, but in doing so they nevertheless pick up and transport pollen to other parts of the forest, assisting the liana even as they're munching through them. I took this modern example as something that likely happened in the deep past, as well. Pollination is not simply a benefit to plants, but part of a relationship that can turn out poorly or even entirely cease. In such a situation, new pollinators might pick up the abandoned roles, the plant might have another way to disperse its pollen, or the plant might go extinct.

Appendix XII

Watching my household felines roll and rub for catnip, I wondered how far into the past the relationship might have gone. I was surprised at the answer. Catnip existed long before tabbies and tortoiseshells, back to when saber-toothed cats still roamed much of the world.

Cats are hypercarnivores. In ecological jargon, that means that they usually prefer to dine on muscle, viscera, and other soft parts of animals. They can strip a skeleton clean but often leave the bones, unlike more adept bone-crackers such as wolves and spotted hyenas. (There are exceptions to this, such as saber-toothed cats that cracked bones more often during hard years at what's now the La Brea asphalt seeps, but these are the sorts of broad-brush distinctions researchers often rely upon.) But anyone who's welcomed a cat into the family knows that felines don't entirely ignore plants. Even if they're not trying to steal your salad from off your fork, cats have a definite fondness for catnip and catmint—both very ancient plants.

Chances are slim that we'll ever know whether saber-toothed cats enjoyed plants with psychoactive compounds as much as my bossy tabby, Strata, at home. Paleontologists would have to find the remains of a saber-toothed cat with preserved gut contents—itself an extreme rarity in the fossil record—and that cat would need to have eaten some catnip soon before death, if any leaves or pollen could be recovered at all. I have a better chance of winning the lottery than finding such a fossil. But I nevertheless believe that saber-toothed cats would have been enthralled by catnip just as living lions, jaguars, and other big cats are. Both cats and catnip have been around for tens of millions of years, and we have every reason to think they would have had some response to catnip just like our purring family members.

I selected *Machairodus* as the feline star of this chapter for a few reasons. Some *Machairodus* species were big— *Machairodus lahayishupup* from North America is estimated to have been about 600 pounds, or about 200 pounds heavier than an adult tiger. *Machairodus* species were also widespread and long-lived, prowling Africa, Eurasia, and North America between 9 and 5.5 million years ago. The window overlaps with when catnip capable of causing a psychoactive reaction was present on Earth.

At present, we know vanishingly little about the social lives of sabercats. Some studies have found that later species such as the famous *Smilodon fatalis* from the La Brea asphalt seeps likely lived in groups, but we don't know the composition of those families or if any other sabercats—which did not live near such a perfect place for creating fossils—were similar. Research on *Smilodon* growth suggests that young cats stayed with their mothers for a long time after birth and formed small family groups. Much like living cats, fossil cats were probably not as solitary as often assumed.

With lions being a notable exception, most big cats have long been assumed to be solo hunters most of the time. Cats don't usually live in large groups like dogs. But that doesn't mean that cats live lonely lives from the time they pounce on their first bug. Among cheetahs, for example, zoologists have found that males from different families will sometimes live and hunt together. Female lions, too, will sometimes form bonds that are more like those of mates than sister huntresses. Evolution thrives on variation, after all, and so cats exhibit a spectrum of different social behaviors that may change through their lives rather than falling into such defined categories as "social" and "gregarious." In this case, I wanted to imagine a pair of sabercats to highlight how catnip can make some of the biggest, scariest-looking cats turn into kittens after a few sniffs.

Evolution is often presented as a constant tinkering process. Traits that are lost are supposedly gone forever and do not return. But botanists have learned that catnip evolved its special qualities twice over its evolutionary history. This would be the equivalent of snakes losing their legs and then regaining them at some later time. And strange as that may sound, as antithetical to the stereotypical image of how evolution works, the reversal makes a lot of sense.

Genetic material is often considered a blueprint for making an organism. But that's not quite right. If DNA is a blueprint, then it's a blueprint that includes instructions for when to tackle certain construction projects and the various possibilities for what *could* happen in a different circumstance. Genes don't dictate so much as offer a range of outcomes that requires input and interaction from the outside world, just as they may change with mutation. Let's apply that to nepetalactone and catnip. The compound originated as a modified form of something the plant was already creating. But the deletion or addition of a gene in just the right spot—for example—could cause catnip to stop producing nepetalactone. Then it's just nip without the cat. Over a long enough timescale, that gene is likely to pick up additional variations as catnip plants reproduced and new mutations were created. In this case, the gene might degrade—a "fossil" gene that suggests what once was, the same way that chickens have busted-up genes related to growing teeth. With a little luck, a mutation could flip that switch back on for nepetalactone production and catnip could start creating it again. From genetic research of modern catnip plants, it seems that's how catnip could have temporarily lost its attractive compound only to regain it.

The coat colors *Machairodus* might have worn offered another chance to talk about the relationships between animals and plants. Pelt patterns of modern cats don't track according to their evolutionary kinship. Lions, tigers, and jaguars are all members of the genus *Panthera*, for example, but their patterns are all strikingly different. The reason is because cat coat colors reflect their habitat and how they hunt, concealment for carnivores that are almost all ambush predators. Many of the world's small cats that live in dense forests and climb trees have long horizontal stripes to help them blend in among the branches and leaves. Cats that live in deserts or grasslands—open spaces with much less cover—tend to be tan, brown, or covered in black polka dot–like spots. Tigers are the only cats in the world with vertical stripes—actually modified spots—and big cats of woodland habitats have their shapes broken up by more open spots and rosettes. It's the plants—how dense or open a habitat is, and how a cat navigates the habitat—that's led to the evolution of different coat colors. Given that *Machairodus* lived during a time of spreading grasslands among patches of woodland, I chose to give them the coats of ambush predators who lurked among the tree trunks.

Appendix XIII

We're not exactly the epitome of arboreal primates. Compared to many living species of ape and monkey, not to mention our fossil relatives dating all the way back to the final days of *T. rex*, our bodies seem far better suited to moving around on the ground. In fact, we have spent decades investigating and arguing about how our ancestors stood upright and how we stand apart from our closest relatives as the apes that spend more time walking and running than scrambling through the branches or traversing the forest floor on all fours. And yet the hallmarks of our arboreal ancestry are still there.

A great deal of our anatomy can be traced through the eons back to life in the trees. It's a little funny to think about as I sit here on my ischia, tapping away at a keyboard. My forward-facing eyes that allow me to focus on the characters popping onto the page and are also able to perceive the distance to the Woods' rose bush across the street? They evolved among tiny Paleocene primates that ran through the canopy and snatched insects for their meals. The dexterous hands that I'm employing to dance across the collection of keys, too, evolved around this time, appendages suited to grasping and grappling, little splints of keratin—nails—serving the purpose far better than claws. But other parts of my makeup are much more recent, at least from an evolutionary standpoint. When I take breaks to play with the soot sprite of a kitten gamboling around my office, dragging toys across the carpet and batting at her as she bats at me, I'm making the most of arm and shoulder flexibility that evolved in the moment just before sweeping changes to global forests gave our ancestors a nudge toward the ground.

One of the key characters of this chapter, *Ardipithecus ramidus*, represents an essential time in human history. One of the earliest known humans—or hominins—*Ardipithecus* seems to shade between our bodies and those of earlier ape ancestors. We can recognize all the same bones, but they are in different shapes. The feet of *Ardipithecus*, for example, are much more ape-like and have a diverging big toe, suitable for stabilizing and holding on while walking over the tops of branches and over the heads of the hyenas, sabercats, and bears below. The early human also lacked the bowl-like hips that would evolve a couple million years later among the australopithecines—"Lucy" and their kin—that help hold all that heavy viscera in place while walking upright. But the arms and shoulders of *Ardipithecus* look quite familiar, with shoulder blades seeming to float behind the rib cage with flexible joints at both the shoulder and the wrist to allow for an incredible range of motion.

If you want to appreciate just how flexible we are, ask your favorite dog to "shake." Dogs have their shoulder blades to the sides of their ribs and evolved to be pursuit hunters, their legs best suited to moving front-to-back. But we can turn our wrists such that our palms can face up or down, left or right, with the ball-and-socket arrangement of our shoulders allowing us to do everything from throw overhand to vogue. It's a gift from our distant ancestors, an anatomical arrangement that allows us—like other primates—to make tools, embrace our little ones to us, and accomplish so much of what we take for granted in a given day.

Ardipithecus didn't move quite like we do. They lived before humans had more or less committed to a life on the ground, around the origins of our genus *Homo*. But they are an important point of interest in our story because of the reason the anatomy of such a human was shunted toward what we now embody rather than beings closer to the chimpanzees, gorillas, and orangutans that we count as our closest living relatives.

Just as the vegetation we dwell in shapes our lives, so does the global climate shift around what plant species grow where and the ever-changing transformation of habitats in the same place through time. Over the course of the Cenozoic, all the time since the asteroid impact 66 million years ago, the world has gone through some dramatic climate changes—including an incredible warming spike about 55 million years ago that was so intense there were alligators living within the Arctic Circle. As we've drawn closer to our present time, however, the global climate has shifted toward being a little cooler and drier, conditions that have favored the spread of grasslands rather than dripping forests. The change has certainly affected our evolutionary family. By about 20 million years ago, woodlands were often interspersed with patches of grasses where early apes foraged. Around 10 million years later, those forests drew back even farther as grasslands spread—the thick forests where apes could easily swing and climb from branch to branch were becoming more restricted toward the tropics. By the time of *Ardipithecus*, these changes were beginning to affect Eastern Africa, too, putting early humans and their great ape relatives in a bit of a bind. The primates could stick to the shrinking forests, hanging on tight, or they could emerge among the more open habitats where hyenas and big cats prowled. Our ancestors were nudged down the latter path, the dangers they faced assisting with their preservation.

We owe the carnivores who ate our prehistoric relatives a debt. Crocodiles, hawks, hyenas, and more, they all preyed on the relatively small and defenseless early humans, often collecting those remains in nests, along lakeshores, or in other places that bones would have a greater chance of preservation. This happenstance is part of the reason why I chose to frame the tale from the perspective of a hungry *Ikelohyaena*, one of the carnivorous mammals that lived in the same riverside habitat as *Ardipithecus*. The second reason is a bit of a narrator's conceit—I wanted to demonstrate the connection between our own bodies and prehistoric plants without trying to get into the heads of the earliest humans directly.

The boundary is arbitrary and drawn on phylogenetic grounds. I don't feel comfortable assigning thoughts and feelings to another human, even one removed from myself by 6 million years. Doing so would feel like I'm undermining the autonomy and experiences of another life that shared enough in common with mine that any assumptions I make will probably create a false image. *Ikelohyaena* is distant enough and different enough from myself that I feel there's an understanding that I cannot possibly know their real feelings, intentions, and experiences outside of what I can glean from the bare bones, and so the fiction is understood by both you and me.

Appendix XIV

Many of the most basic questions in paleontology are the most challenging to answer. We often receive the specifics of a particular species first, then work outward to the broader ways ecosystems and entire biomes have functioned across time. So when a friend asked me, "When was there the first fall?" I realized that I had no idea.

If we think about it broadly, autumn has existed for a very long time. The tilt of the Earth and its orbital path around the sun have created seasonal shifts between the long days of summer and the chilly, extended evenings of winter for millions of years. When paleontologists think about dinosaurs that lived in prehistoric Alaska around 70 million years ago, for example, they have to account for the fact that the reptiles needed to cope with months of darkness in polar winters dictated by the bare mechanics of how our planet moves through space. But part of why we call autumn "fall" is because of the falling leaves, and when that seasonal change started to happen is more difficult to discern.

We often think of angiosperms when fall starts to settle in, but flowering plants aren't the only ones to change around this time. The leaves of modern ginkgoes, for example, turn golden in the fall before lilting to the ground. Given that ginkgoes were much more widespread and diverse through the Mesozoic, it's possible that autumn color changes go back to the heyday of the dinosaurs. Then again, we only have one modern tree to go on, and color changes in living ginkgoes appears to be linked to temperature. The Mesozoic world was, on average, warmer than our current interglacial, so it might be that ginkgoes stayed green through a seemingly endless summer.

Following fall through fossil angiosperms is also a complicated task. Future discoveries may soon change the picture, and some research has already suggested that flowering plants evolved much earlier in the Mesozoic, but the oldest definitive evidence of angiosperms goes back to about 125 million years ago. Even then, conifers, cycads, ginkgoes, and other plants were far more prevalent on the landscape until the close of the Cretaceous—it was the asteroid impact 66 million years ago that cleared the ground for flowering plants to shift the balance. If you could visit the Earth during an autumn month—depending on the hemisphere—in that window of 59 million years, it's unlikely you'd see entire forests blazing with color like we do now. Even among relatively familiar plants that have roots in the Cretaceous, like magnolias, we don't know for sure whether they changed colors like their modern counterparts do.

I could have made the case that Cretaceous plants changed color just by virtue of trees evolving to fit the circumstances of our planet. Daylight and temperatures change with autumn as phenomena extending from what makes Earth so unique. Doing so felt like a reach without something more specific to pin it on, but fortunately paleobotanists have indeed found bright autumn leaves in the fossil record. They're just more recent to us in time than *Triceratops*.

About 20 miles north of Göttingen, Germany, in Willershausen, there's a clay pit that preserves remnants from the organisms that lived in the area during the Pliocene, or about 3 million years ago. Paleontologists eventually gathered various parts of the ancient ecosystem, from microscopic diatoms to termites and the extinct elephant *Anancus*. But the plants were really what made the site famous. The fossil leaves found in the clay pit were gorgeous and abundant— one researcher collected more than 30,000 plant specimens from this one place. Some of the plants even seemed to retain their original colors. Geologists would later learn that the various fossils had accumulated in what once had been a lake, one with extremely salty bottom waters. The brine on the lake bottom is what allowed so many plants to be so well preserved, including leaves that had been changing shades for the fall.

The hues of a fossil bone or plant leaf are often very different from the color of the living organism. When living things become permineralized, minerals transported by water percolate through the tissues and replace them to varying degrees. This is part of why dinosaur skeletons—even of the same species—are often different colors thanks to different minerals involved in the mineralization process. The same goes for fossil leaves, which may look like one color to the naked eye but not represent the original shade. But in the case of the Willershausen fossils, paleobotanists have been able to use fluorescence to detect the original shades.

UV light is an important tool for paleontologists. Among vertebrate fossils, UV light can help paleontologists detect fakes—as plaster or other materials will look different from actual bone under the light—and identify otherwise invisible bits of anatomy like wispy feathers. In 2021, paleobotanists applied the same technique to plants from Willershausen. Fossils of ash, box, and ivy leaves look tan under white light, but under fluorescent UV exposure the fossils show up in shades of red, orange, and highlighter yellow. Not all the leaves from the site fluoresced strongly, but many did—the colors the leaves wore when they fell into the lake remained in their fossilized tissues. The difference between the colors was because of chlorophyll degradation, or when plants absorb their green chlorophyll in the autumn and the red, orange, and yellow carotenoids become more apparent. By using UV light, paleobotanists were able to see shades that prehistoric Germany's plants wore around 3 million years ago.

While future studies will have to check, it's likely that the record of autumn leaves in the fossil record goes back even further. The study on the Willershausen fossils noted that there are sites over 47 million years old that fluoresce in colors from red to green under UV light. Finding the right plants for such investigations is a tricky task. Paleontologists have the best chance with leaves that were buried quickly in fine-grained sediments, in water with little oxygen to slow the decay process. Nevertheless, now that paleobotanists know how to look for the right clues, experts can track autumn blazes through Deep Time.

Appendix XV

There are some fossils I simply have a soft spot for, and many of the prehistoric animals reconstructed in Manhattan's American Museum of Natural History are among them. The museum was one of my favorite places to visit during my college years, when I'd take the train from New Brunswick, New Jersey, to Penn Station followed by a quick C train run uptown to the museum's own subway stop. And while there was always plenty to see and ponder in those classic halls, I could never leave without visiting the Warren mastodon.

The Warren mastodon still stands in the museum's Hall of Advanced Mammals, a silly designation that carries a whiff of hierarchy to it. But I'm glad the old beast is still there. The skeleton was originally found in Newburgh, New York, in 1845, the Ice Age bones stuck in a bog where the mastodon likely died. There were no professional paleontologists or field experts at that time. When a mastodon was found, the effort was simply to dig in and get the bones out—often for display in small or private museums. All of which is to say that we don't have detailed notes on the excavation of the Warren mastodon, but it's said that the skeleton was found in a standing position with its head thrown back and front legs seeming to scramble for purchase. Perhaps, paleontologists have proposed, the mastodon got stuck and helplessly sank into the bog that preserved it. With the exception of a few bones from the tail and toes, the entire mastodon was preserved—right down to the gut contents.

Paleontologists have found better-preserved gut contents from desiccated mammoth carcasses in Siberia and ancient mammoth plops that contain remnants of mushrooms that grew in the green pats. Some giant sloths have been found with their gut contents, too, preserved for thousands of years in dry caves seemingly overflowing with partially digested plant material. Any of these beasts would have made a suitable case study for the last chapter, but I couldn't resist going with a sentimental favorite. I've always adored mastodons, beasts that thrived in the interglacials and, given that we are in such a time now, should still be here.

The title could have gone to some other ancient creature, but the American mastodon—*Mammut Americanum*—may be the most important extinct animal ever discovered. Paleontology wouldn't be the same without it. Indigenous peoples had recognized the abundant bones of mastodons as the remains of once-living animals before scientists ever saw so much a molar, and enslaved Black people abducted to America noticed the resemblance between mastodon bones and elephant remains before naturalists did. When white naturalists finally caught on, they debated the nature of what this "American Incognitum" could be. Working from bits and pieces, some believed that the beast was a predator who leapt onto deer and other smaller prey to rip them apart with its rough molars. Others wondered if the creature might still dwell in the interior of the continent—Thomas Jefferson, who struggled to accept the reality of extinction, famously instructed the Lewis and Clark expedition to keep an eye out for the mastodon and other giant beasts during the foray's colonialist mission to reach the Pacific. But France was the center of anatomical science in the West during the late eighteenth century, and it wasn't until 1799 that naturalist Georges Cuvier concluded that what we now call the American mastodon was a definitively extinct giant that disappeared from the planet save for its bones. The fact that the mastodon resembled living elephants—but was clearly and consistently different, with no place to hide during an age of expansive colonialism—allowed experts to finally accept the basic fact that animals even *could* go extinct and that the fossil record contained the remnants of times very different from what we see around us now.

Some of the early mastodon finds have been lost. Naturalists and curio collectors of centuries past had little idea how to care for and preserve fossils. Even in the case of the Warren mastodon itself, the elephant's tusks began to crumble soon after it was put on display—they were replaced with papier-mâché replicas until paleontologists later pieced the originals back together. But there was more to the mastodon than just the bones. Preserved back near the hips, in what would have been the animal's hind gut, were the shredded and digested tatters of what the mastodon had recently been eating.

Gut contents can never give us the total picture of an animal's diet. Food might vary by season and by year, so what we find in a fossil skeleton really only represents what an animal had recently dined on. Nevertheless, the gut contents found with the Warren mastodon confirmed that the animal ate plants—not meat, as earlier natural history aficionados thought—and eventually served as important comparative material for a massive deposit of mastodon dung found in a Florida spring.

While they wouldn't pass modern paleontological standards, original reports of the Warren mastodon remains mentioned about "five to six bushels" of swollen twigs and branches in a cylinder-like shape against the pelvis—more or less the shape of elephant intestines. The vegetation was hemlock, not chewed so much as busted by the cusps of the mastodon's molars. Other historic mastodon finds reportedly had clumps of plant material within the confines of their skeletons, too, often the branches of conifers and other trees. The repeated signal was clear, that mastodons preferred to browse on tree branches that they barely chewed before swallowing and letting their vast fermenting organs do their work. All of this material allowed paleontologists to better interpret the treasure trove of mastodon dung found in Florida, indicating that the herbivores ate wild grapes, acorns, persimmon, hawthorn, plum, and cypress branches, altogether making them very influential megaherbivores that changed the habitat by what they ate.

In addition to considering the mastodon's role in shaping the forest and an in-between step by which broken-down plants affected the climate, I wanted to make sure to highlight other lives that were part of the same forests. Leafcutter bees were well established by the latter part of the Ice Age, and fossil leafcutter offspring have been found in the asphalt seep of La Brea in California. Research at that site, as well, has helped outline the dramatic changes North America's ecosystems were going through toward the end of the Pleistocene. We live in a world still filled with Ice Age survivors, but we are missing some of the most characteristic species, and so I wanted to include this new research to help explain the combination of factors that altered the recent past. The inclusion of *Monotropa*, as well, has a personal connection. I had never heard of the little plants before I noticed them on a hike through a New York forest. The plants evolved to live on fungi that had already established their own close relationship with plants, and genetic studies indicate that *Monotropa* withdrew to warmer "refugia" during times of ice while expanding again as the global climate warmed. Just as mastodons wandered away from the ice during the great glaciations, *Monotropa* pulled back, too, only to return.

I wanted to write this chapter as a bridge to Pando in the conclusion, a survivor from the Ice Age. In doing so, I couldn't avoid the reality that I'm imagining landscapes that have their own histories and connections to people who saw these forests as they grew. The earliest humans evolved in eastern Africa by 6 million years ago, spinning off an array of species that paleoanthropologists are still in the process of discovering. Our own species, *Homo sapiens*, goes back to about 300,000 years ago, and our ancestors very quickly began to spread around the world. In the context of this chapter's scene in Pleistocene New York, people arrived in North America by at least 21,000 years ago—although this is likely a gross underestimate. Archaeologist Paulette Steeves notes that people may have been in the Americas by 130,000 years ago, the seeming gap between the earliest evidence and the actual date having as much to do with archaeology's distorted relationship with Indigenous peoples as the nature of the historic record. All of which is to say, what I am presenting in this chapter is conjecture based on scientific research when there were certainly people who witnessed *Mammut* and the landscapes of ancient New York firsthand. Whatever I imagine cannot match what they experienced.

I can look at the bone of a mastodon and see the life of an

animal that lived long ago. My entire career has been based around this warm, curious relationship with fossils and what they might tell us—not just about the history of life but those individual lives that sparked and vanished during eras when my own existence was nothing more than a distant and remote possibility. But the way I see a fossil and what I feel that bone represents is not the only way to approach the past, nor should it be. My hope is that I've introduced you to enough points of contact and ancient visions that you've developed your own curiosity and questions in the process, perspectives that are different from mine. What I have laid out in these pages is ancient life as I see it. When you think of these times, these lives, what do you see?

Acknowledgments

The shape of the book I'm writing molds the way I think about the project. When I wrote about bones, I thought about my explanations and turns of phrase as the soft tissue on an osteological framework of core ideas. Writing about the deep and entangled relationship of animal and plant life on our planet, I've thought more in terms of ecosystems—the rainbow of influences both seen and invisible that results in something uniquely vibrant. These acknowledgments are my attempt at honoring those unseen factors that pruned and fostered this book. And what better day to complete my musings than during the cusp of spring, when snowdrops and many-colored crocus are pushing their way out of the ground and trees take on a red haze from buds preparing to unfurl?

My agent, Deirdre Mullane, has been a champion of my work since *Skeleton Keys*. She's always believed in my work, even when I've caused her a headache or two. You once called me a "force of nature," Deirdre, and I'm doing my best to live up to the reputation.

I'll admit that I've never been the best gardener, even of my own words. I'm grateful to Cassidy Graham of St. Martin's for not only believing in this project from the first whisper of its possibility, but for spotting all the places where I needed to prune, weed, and let some parts of the story grow a little wilder. Thank you, Cassidy, for your patience and guidance as I've tried to think my way through all the book's tangled topics.

Kory Bing's artistic efforts for this book were nothing less than heroic. It's not easy depicting such a broad range of living things, some of which are rarely illustrated even as references. Thank you, Kory, for bringing so many scenes in this book and *The Last Days of the Dinosaurs* to life.

I'm grateful to the many friends who knew well enough not to ask me, "How's the book going?" during the process but always listened warmly as I gathered my thoughts about the stories I wanted to tell or had some new tidbit of information I was excited to tell them about. Blue Neustifter, Bee Brookshire, Alex Porpora, Carrie Levitt-Bussian, the Monday night Flashback Discord group, and others have all seen me through the process, many of them for the umpteenth time. I appreciated their reminders to be kind to myself as I struggled to find just the right words.

Of course, it's impossible to write a book without self-doubt. The narrative lives in my head, and on my hard drive, for months, something that can't be taken in over a moment or even an hour. It's intimidating. But my partner, Splash, has never once let me believe that I can't find some way to express the stories I so desperately want to tell. Many of the ideas in this book can be traced to ideas I've batted back and forth as we've sipped lattes on walks through shade-covered city streets, stopping to look at the flowers, bees, spiders, birds, and trees. It's been a joy to share the process with someone who loves to flip over rocks and gets excited when there's a woodpecker in the yard like I do, and I hope I've written a fitting tribute to those affections.

Not all my family members read my work. My loving cats Hobbes, Joey, and Strata, as well as Jet the German shepherd, could not care even a little about my books. And that's why I've appreciated their unconditional affection, all the times they've reminded me it's time for a walk, it's time to cuddle, or it's time to rest. Even as I type this, Joey and Strata are cuddled together on my legs as Hobbes dozes next to me and Jet twitches, chasing after whatever is in his doggy dreams from his favorite spot on the couch. I also had to say good-bye to feline family—Teddy and Terra—during the process of writing this book, and as much as I miss them I'm glad we got any time at all. Losing them, but welcoming Strata and Joey home, undoubtedly affected the threads of death and new life that run through some of these chapters.

And though it's likely that we've never met, thank you for picking up this book. So much of life is fragmented, much of it constantly vying for our attention as important or essential. Just as with *The Last Days of the Dinosaurs*, my hope for this book is resonance—that you have found something in these pages that unlocks or alters the way you experience the world and all the wondrous life that's entangled on it even in this moment. Thank you for hiking along with me a while.

References

1. Sex in the Shallows

J. Archibald. 2009. "Secondary Endosymbiosis." In *Encyclopedia of Microbiology*. 3rd ed., edited by M. Schaechter. Waltham, MA: Academic Press.

A. Bowles, C. Williamson, T. Williams, et al. 2023. "The origin and early evolution of plants." *Trends in Plant Science* 28 (3): 312–329.

N. Butterfield. 2000. "*Bangiomorpha pubescens* n. gen., n. sp:: implications for the evolution of sex, multicellularity, and the Mesoproterozoic/Neoproterozoic radiation of eukaryotes." *Paleobiology* 26 (3): 386–404.

D. Canfield, L. Ngombi-Pemba, E. Hammarlund, et al. 2013. "Oxygen dynamics in the aftermath of the Great Oxidation of Earth's atmosphere." *PNAS* 110 (42): 16736–16741.

L. Falcón, S. Magallón, A. Castillo. 2010. "Dating the cyanobacterial ancestor of the chloroplast." *The ISME Journal* 4: 777–783.

S. Garg, V. Zimorski, W. Martin. 2016. "Endosymbiotic Theory." In *Encyclopedia of Evolutionary Biology*, edited by R. Kliman. Waltham, MA: Academic Press.

T. Gibson, P. Shih, V. Cumming, et al. 2018. "Precise age of *Bangiomorpha pubescens* dates the origin of eukaryotic photosynthesis." *Geology* 46 (2): 135–138.

L. Graham, M. Cook, J. Busse. 2000. "The origin of plants: body plan changes contributing to a major evolutionary radiation." *PNAS* 97 (9): 4535–4540.

P. Keeling. 2010. "The endosymbiotic origin, diversification and fate of plastids." *Philosophical Transactions of the Royal Society B: Biological Sciences* 365 (1541): 729–748.

J. Olejarz, Y. Iwasa, A. Knoll, M. Nowak. 2021. "The Great Oxygenation Event as a consequence of ecological dynamics modulated by planetary change." *Nature Communications* 12 (3985).

J. Olson. 2006. "Photosynthesis in the Archean Era." *Photosynthesis Research* 88: 109–117.

B. Sabater. 2018. "Evolution and function of the chloroplast. Current investigations and perspectives." *International Journal of Molecular Sciences* 19 (10): 3095.

2. Worts and All

D. Edwards, P. Selden, J. Richardson, L. Axe. 1995. "Coprolites as evidence for plant-animal interaction in Siluro-Devonian terrestrial ecosystems." *Nature* 377: 329–331.

P. Gensel. 2008. "The earliest land plants." *Annual Review of Ecology, Evolution and Systematics* 39: 459–477.

T. Lenton, M. Crouch, M. Johnson. 2012. "First plants cooled the Ordovician." *Nature Geoscience* 5: 86–89.

T. Lenton, T. Dahl, S. Daines, et al. 2016. "Earliest land plants created modern levels of atmospheric oxygen." *PNAS* 113 (35): 9704–9709.

M. Libertín, J. Kvaček, J. Bek, et al. 2018. "Sporophytes of polysporangiate land plants from the early Silurian period may have been photosynthetically autonomous." *Nature Plants* 4: 269–271.

R. MacNaughton, J. Cole, R. Dalrymple, et al. 2002. "First steps on land: Arthropod trackways in Cambrian-Ordovician eolian sandstone, southeastern Ontario, Canada." *Geology* 30 (5): 391–394.

M. Marshall. 2012. "First land plants plunged Earth into ice age." *New Scientist*. February 1, 2012. https://www.newscientist.com/article/dn21417-first-land-plants-plunged-earth-into-ice-age/.

J. Morris, M. Puttick, J. Clark. 2018. "The timescale of early land plant evolution." *PNAS* 115 (10): E2274–E2283.

E. Pennisi. 2018. "Land plants arose earlier than thought—and may have had a bigger impact on the evolution of animals." *Science*. February 19, 2018. https://www.science.org/content/article/land-plants-arose-earlier-thought-and-may-have-had-bigger-impact-evolution-animals.

C. Wellman, P. Osterloff, U. Mohiuddin. 2003. "Fragments of the earliest land plants." *Nature* 425: 282–285.

J. Yang, T. Lan, X. Zhang, M. Smith. 2023. "*Protomelission* is an early dasyclad alga and not a Cambrian bryozoan." *Nature* 615: 468–471.

M. Zhao, B. Mills, W. Homoky, C. Peacock. 2023. "Oxygenation of the Earth aided by mineral-organic carbon preservation." *Nature Geoscience* 16: 262–267.

W. Zhao, X. Zhang, G. Jia, et al. 2021. "The Silurian-Devonian boundary in

East Yunnan (South China) and the minimum constraint for the lungfish-tetrapod split." *Science China Earth Sciences* 64: 1784–1797.

3. The Forest Primeval

N. Davies, R. Garwood, W. McMahon, et al. 2021. "The largest arthropod in Earth history: insights from newly discovered *Arthropleura* remains (Serpukhovian Stainmore Formation, Northumberland, England)." *Journal of the Geological Society* 179 (3): 2021–115.

L. Labeeuw, P. Martone, Y. Boucher, R. Case. 2015. "Ancient origin of the biosynthesis of lignin precursors." *Biology Direct* 10 (23).

A. Mann, A. Henrici, H. Sues, S. Pierce. 2023. "A new Carboniferous edaphosaurid and the origin of herbivory in mammal forerunners." *Scientific Reports* 13 (4459).

M. Nelsen, W. DiMichele, S. Peters, C. Boyce. 2016. "Delayed fungal evolution did not cause the Paleozoic peak in coal production." *PNAS* 113 (9): 2442–2447.

W. Verberk, D. Bilton. 2011. "Can oxygen set thermal limits in an insect and drive gigantism?" *PLOS ONE* 6 (7): e22610.

4. Fire and Water

S. Ash, G. Creber. 2003. "The Late Triassic *Araucarioxylon arizonicum* trees of the Petrified Forest National Park, Arizona, USA." *Palaeontology* 43 (1): 15–28.

V. Baranyi, T. Reichgelt, P. Olsen, et al. 2018. "Norian vegetation history and related environmental changes: new data from the Chinle Formation, Petrified Forest National Park (Arizona, SW USA)." *GSA Bulletin* 130 (5–6): 775–795.

B. Byers, L. DeSoto, D. Chaney. 2020. "Fire-scarred tree from the Late Triassic shows a pre-fire drought signal." *Scientific Reports* 10: 20104.

G. Mustoe. 2023. "Silification of wood: an overview." *Minerals* 13 (2): 206.

M. Qvarnström, M. Fikáček, J. Wernström, et al. 2021. "Exceptionally preserved beetles in a Triassic coprolite of putative dinosauriform origin." *Current Biology* 31 (15): 3374–3381.e5.

K. Zeigler. 2003. "Taphonomic analysis of the Snyder Quarry: a fire-related upper Triassic vertebrate fossil assemblage from north-central New Mexico." *New Mexico Museum of Natural History and Science Bulletin* 24: 49–62.

5. Land of Giants

P. Del Tredici. 2008. "Wake up and smell the ginkgos." *Arnoldia* 66 (2).

G. Engelmann, D. Chure, A. Fiorillo. 2004. "The implications of a dry climate for the paleoecology of the fauna of the Upper Jurassic Morrison Formation." *Sedimentary Geology* 167 (3–4): 297–308.

F. Gill, J. Hummel, A. Sharifi, et al. 2018. "Diets of giants: the nutritional value of sauropod diet during the Mesozoic." *Palaeontology* 61 (5): 647–658.

M. Howell, C. Gee, C. Böttger, K. Südekum. 2023. "Digestibility of dinosaur food plants revisit and expanded: previous data, new taxa, microbe donors, foliage maturity, and seasonality." *PLOS ONE* 18 (12): e0291058.

J. Hummel, C. Gee, K. Südekum, et al. 2008. "*In vitro* digestibility of fern and gymnosperm foliage: implications for sauropod feeding ecology and diet selection." *Proceedings of the Royal Society B* 275 (1638): 1015–1021.

J. Parrish, F. Peterson, C. Turner. 2004. "Jurassic 'savannah'—plant taphonomy and climate of the Morrison Formation (Upper Jurassic, Western USA)." *Sedimentary Geology* 167 (3–4): 137–162.

6. In Bloom

R. Bateman. 2020. "Hunting the snark: the flawed search for the mythical Jurassic angiosperms." *Journal of Experimental Biology* 71 (1): 22–25.

E. Friis, J. Doyle, P. Endress, Q. Leng. 2003. "*Archaefructus*—angiosperm precursor or specialized early angiosperm?" *Trends in Plant Science* 8 (8): P369–373.

G. Kozlowski, M. Stoffel, S. Bétrisey. 2014. "Hydrophobia of gymnosperms: myth or reality? A global analysis." *Ecohydrology* 8 (1): 105–112.

J. Qiang, L. Honqi, M. Bowe, et al. 2010. "Early Cretaceous *Archaefructus* sp. nov. with bisexual flowers from Beipiao, Western Liaoning, China." *Acta Geologica Sinica* 78 (4): 883–892.

X. Wang, X. Xu. 2011. "A new iguanodontid (*Jinzhousaurus yangi* gen. et ap. nov.) from the Yixian Formation of western Liaoning, China." *Chinese Science Bulletin* 46: 1669–1672.

Y. Wu, Y. Ge, H. Hu, et al. 2023. "Intra-gastic phytoliths provide evidence for folivory in basal avialans of the Early Cretaceous Jehol Biota." *Nature Communications* 14: 4558.

7. A Sticky Situation

A. Borkent, D. Grimaldi. 2004. "The earliest fossil mosquito (Diptera: Culicidae), in Mid-Cretaceous Burmese Amber." *Annals of the Entomological Society of America* 97 (5): 882–888.

J. Daza, E. Stanley, A. Bolet, et al. 2020. "Enigmatic amphibians in mid-

Cretaceous amber were chameleon-like ballistic feeders." *Science* 370 (6517): 687–691.

X. Delclòs, E. Peñalver, E. Barrón. 2023. "Amber and the Cretaceous Resinous Interval." *Earth-Science Reviews* 104486.

H. Jiang, F. Tomaschek, A. Muscente, et al. 2022. "Widespread mineralization of soft-bodied insects in Cretaceous amber." *Geobiology* 20 (3): 363–376.

X. Martínez-Delclòs, D. Briggs, E. Peñalver. 2004. "Taphonomy of insects in carbonates and amber." *Palaeogeography, Palaeoclimatology, Palaeoecology* 203 (1–2):19–64.

S. Trapp, R. Croteau. 2001. "Defensive resin biosynthesis in conifers." *Annual Review of Plant Physiology and Plant Molecular Biology* 52: 689–724.

C. Vázquez-González, R. Zas, N. Erbilgin, et al. 2020. "Resin ducts as resistance traits in conifers: linking dendrochronology and resin-based defences." *Tree Physiology* 40 (1): 1313–1326.

8. Rainforests and Revival

M. Carvalho, C. Jaramillo, F. La Parra. 2021. "Extinction at the end-Cretaceous and the origin of modern Neotropical rainforests." *Science* 372 (6537): 63–68.

P. Davis, S. Caylor, C. Whippo, R. Hangarter. 2011. "Changes in leaf optical properties associated with light-dependent chloroplast movements." *Plant, Cell & Environment* 34 (12): 2047–2059.

J. Head, J. Bloch, A. Hastings, et al. 2009. "Giant boid snake from the Palaeocene neotropics reveals hotter past equatorial temperatures." *Nature* 475: 715–717.

G. Nawkar, M. Legris, A. Goyal, et al. 2023. "Air channels create a directional light signal to regulate hypocotyl phototropism." *Science* 382 (6673): 935–940.

L. Weaver, H. Fulghum, D. Grossnickle, et al. 2022. "Multituberculate mammals show evidence of a life history strategy similar to that of placentals, not marsupials." *The American Naturalist* 220 (3).

S. Wing, F. Herrera, C. Jaramillo, et al. 2009. "Late Paleocene fossils from the Cerrejón Formation, Colombia, are the earliest record of Neotropical rainforest." *PNAS* 106 (44): 18627–18632.

9. Adrift

M. Bond, M. Tejedor, K. Campbell Jr., et al. 2015. "Eocene primates of South America and the African origins of New World monkeys." *Nature* 520: 538–541.

R. Calsbeek, T. Smith. 2003. "Ocean currents mediate evolution in island lizards." *Nature* 426: 552–555.

F. de Oliveira, E. Molina, G. Marroig. 2009. "Paleogeography of the South Atlantic: a route for primates and rodents into the New World?" In *South American Primates*, P. Garber et al. New York: Springer.

T. Gallaher, M. Callmander, S. Buerki, S. Keeley. 2015. "A long distance dispersal hypothesis for the Pandanaceae and the origins of the *Pandanus tectorius* complex." *Molecular Phylogenetics and Evolution* 83: 20–32.

L. Hautier, R. Sarr, R. Tabuce, et al. 2012. "First prorastomid sirenian from Senegal (Western Africa) and the Old World origin of sea cows." *Journal of Vertebrate Paleontology* 23 (5): 1218–1222.

L. Marivaux, F. Negri, P. Antoine, et al. 2023. "An eosimiid primate of South Asian affinities in the Paleogene of Western Amazonia and the origin of New World monkeys." *PNAS* 120 (28): e2301338120.

R. Savage, D. Domning, J. Thewissen. 1994. "Fossil sirenia of the west Atlantic and Caribbean region. V. The most primitive known sirenian, *Prorastomus sirenoides* Owen, 1855." *Journal of Vertebrate Paleontology* 14 (3): 427–449.

E. Seiffert, M. Tejedor, J. Fleagle. 2020. "A parapithecid stem anthropoid of African origin in the Paleogene of South America." *Science* 368 (6487): 194–197.

10. SEAS OF GRASS

H. Appel, R. Cocoroft. 2014. "Plants respond to leaf vibrations caused by insect herbivore chewing." *Oecologia* 175 (4): 1257–1266.

G. Boardman, R. Secord. 2013. "Stable isotope paleoecology of White River ungulates during the Eocene-Oligocene transition in northwestern Nebraska." *Palaeogeography, Palaeoclimatology, Palaeoecology* 375: 38–49.

R. Gillham. 2019. "Changes in mammalian abundance through the Eocene-Oligocene climate transition in the White River Group of Nebraska, USA." Department of Earth and Atmospheric Sciences, University of Nebraska–Lincoln.

E. Gustafson. 1986. "Preliminary biostratigraphy of the White River Group (Oligocene Chadron and Brule Formations) in the vicinity of Chadron, Nebraska." *Transactions of the Nebraska Academy of Sciences and Affiliated Societies* 209: 7–19.

A. Mithöfer, M. Maffei. 2016. "General mechanisms of plant defense and plant toxins." In *Plant Toxins*, edited by C. Carlini, R. Ligabue-Brain. Dordrecht, Netherlands: Springer.

O. Sanisidro, M. Mihlbachler, J. Cantalapiedra. 2023. "A macroevolutionary pathway to megaherbivory." *Science* 380 (6645): 616–618.

11. Partners in Pollination

T. Bao, B. Wang, J. Li, D. Dilcher. 2019. "Pollination of Cretaceous flowers." *PNAS* 116 (49): 24707–24711.

S. Cappellari, H. Schaefer, C. Davis. 2013. "Evolution: pollen or pollinators—which came first?" *Current Biology* 23 (8): R316–R318.

M. Daniel. 1976. "Feeding by the short-tailed bat (*Mystacina tuberculate*) on fruit and possibly nectar." *New Zealand Journal of Zoology* 3 (4): 391–398.

T. Fleming, C. Geiselman, W. Kress. 2009. "The evolution of bat pollination: a phylogenetic perspective." *Annals of Botany* 104 (6): 1017–1043.

S. Hand, D. Lee, T. Worthy, et al. 2015. "Miocene fossils reveal ancient roots for New Zealand's endemic *Mystacina* (Chiroptera) and its rainforest habitat." *PLOS ONE* 10 (6): e0128871.

E. Heithaus. 1982. "Coevolution between bats and plants." In *Ecology of Bats*, edited by T. Kunz. Boston: Springer.

A. Khramov, T. Foraponova, P. Węgierek. 2023. "The earliest pollen-loaded insects from the Lower Permian of Russia." *Biology Letters* 19 (3).

J. Lord. 1991. "Pollination and seed dispersal in *Freycinetia baeuriana*, a dioecious liane that has lost its bat pollinator." *New Zealand Journal of Botany* 29 (1): 83–86.

G. Mayr, V. Wilde. 2014. "Eocene fossil is earliest evidence of flower-visiting by birds." *Biology Letters* 10 (5).

C. Peña-Kairath, X. Delclòs, S. Álvarez-Parra, et al. 2023. "Insect pollination in deep time." *Trends in Ecology and Evolution* 38 (8): 749–759.

E. Peñalver, C. Labandeira, E. Barrón, et al. 2012. "Thrips pollination of Mesozoic gymnosperms." *PNAS* 109 (22): 8623–8628.

12. 'Nip Trip

W. Allen, I. Cuthill, N. Scott-Samuel, R. Baddeley. 2010. "Why the leopard got its spots: relating pattern development to ecology in felids." *Proceedings of the Royal Society B* 278 (1710).

C. Janis, J. Damuth, J. Theodor. 2022. "The origins and evolution of the North America grassland biome: the story from hoofed mammals." *Palaeogeography, Palaeoclimatology, Palaeoecology* 177 (1–2): 183–198.

C. Kaelin, X. Xu, L. Hong, et al. 2012. "Specifying and sustaining pigmentation patterns in domestic and wild cats." *Science* 337 (6101): 1536–1541.

N. Melo, M. Capek, O. Arenas, et al. 2021. "The irritant receptor TRPA1 mediates the mosquito repellant effect of catnip." *Current Biology* 31 (9): 1988–1994.E5.

J. Orcutt, J. Calede. 2021. "Quantitative analyses of feliform humeri reveal the existence of a very large cat in North America during the Miocene." *Journal of Mammalian Evolution* 28: 729–751.

A. Ortolani. 2008. "Spots, stripes, tail tips and dark eyes: predicting the function of carnivore colour patterns using the comparative method." *Biological Journal of the Linnean Society* 67 (4): 433–476.

13. Far from the Tree

T. Cerling, J. Wynn, S. Anganje, et al. 2011. "Woody cover and hominin environments in the past 6 million years." *Nature* 476: 51–56.

S. Elton. 2008. "The environmental context of human evolutionary history in Eurasia and Africa." *Journal of Anatomy* 212 (4): 377–393.

M. Gani, N. Gani. 2011. "River-margin habitat of *Ardipithecus ramidus* at Aramis, Ethiopia 4.4 million years ago." *Nature Communications* 2 (602).

C. Gilbert, K. Pugh, J. Fleagle. 2020. "Dispersal of Miocene hominoids (and pliopithecoids) from Africa to Eurasia in light of changing tectonics and climate." In *Biological Consequences of Plate Tectonics*, edited by G. Prasad, R. Patnaik. Cham, Switzerland: Springer.

C. Gibert, A. Vignoles, C. Contoux, et al. 2022. "Climate-inferred distribution estimates of mid-to-late Pliocene hominins." *Global and Planetary Change* 210: 103756.

A. Louchart, H. Wesselman, R. Blumenschine, et al. 2009. "Taphonomic, avian, and small-vertebrate indicators of *Ardipithecus ramidus* habitat." *Science* 326 (5949): 66–66e4.

M. Maslin, C. Brierly, A. Milner, et al. 2014. "East African climate pulses and early human evolution." *Quaternary Science Reviews* 101: 1–17.

M. Trauth, M. Maslin, A. Deino, et al. 2007. "High- and low-latitude forcing of Plio-Pleistocene East African climate and human evolution." *Journal of Human Evolution* 53 (5): 475–486.

14. Ancient Autumn

D. Ferguson, E. Knobloch. 1998. "A fresh look at the rich assemblage from the Pliocene sink-hole of Willershausen, Germany." *Review of Palaeobotany and Palynology* 101 (1–4): 271–286.

R. Rivals, D. Mol, F. Lacombat, et al. 2015. "Resource partitioning and niche separation between mammoths (*Mammuthus rumanus* and *Mammuthus meridionalis*) and gomphotheres (*Anancus arvernensis*) in the Early Pleistocene of Europe." *Quaternary International* 379: 164–170.

T. Schmidt-Schultz, M. Reich, M. Schultz. 2021. "Exceptionally preserved extracellular bone matrix proteins from the late Neogene probosciedean *Anancus* (Mammalia: Proboscidea)." *PalZ* 95: 757–765.

K. Wolkenstein, G. Arp. 2021. "Taxon- and senescence-specific fluorescence of colored leaves from the Pliocene Willershausen Lagerstätte, Germany." *PalZ* 95: 747–756.

A. Zazzo, H. Bocherens, D. Billiou, et al. 2000. "Herbivore paleodiet and paleoenvironmental changes in Chad during the Pliocene using stable isotope ratios of tooth enamel carbonate." *Paleobiology* 26 (2): 294–309.

15. AFTER THE ICE

E. Bakker, J. Gill, C. Johnson, et al. 2015. "Combining paleo-data and modern exclosure experiments to assess the impact of megafauna extinctions on woody vegetation." *PNAS* 113 (4): 847–855.

G. Beatty, J. Provan. 2011. "Phylogeographic analysis of North American populations of the parasitic herbaceous plant *Monotropa hypopitys* L. reveals a complex history of range expansion from multiple late glacial refugia." *Journal of Biogeography* 38 (8): 1585–1599.

A. Holden, J. Koch, T. Griswold, et al. 2014. "Leafcutter bee nests and pupae from the Rancho La Brea Tar Pits of Southern California: implications for understanding the paleoenvironment of the Late Pleistocene." *PLOS ONE* 9 (4): e94724.

Y. Malhi, C. Doughty, M. Galetti, et al. 2016. "Megafauna and ecosystem function from the Pleistocene to the Anthropocene." *PNAS* 113 (4): 838–846.

L. Newsom, M. Mihlbachler. 2006. "Mastodons (*Mammut Americanum*) diet foraging patterns based on analysis of dung deposits." In *First Floridians and Last Mastodons: The Page-Ladson Site in the Aucilla River*, edited by S. Webb. Dordrecht, Netherlands: Springer.

F. O'Keefe, R. Dunn, E. Weitzel, et al. 2023. "Pre-Younger Dryas megafauna extirpation at Rancho La Brea linked to fire-driven state shift." *Science* 381 (6659).

F. Smith, J. Hammond, M. Balk, et al. 2015. "Exploring the influence of ancient and historic megaherbivore extirpations on the global methane budget." *PNAS* 113 (4): 874–879.

About the Author

RILEY BLACK (she/they) is the award-winning author of *Skeleton Keys, My Beloved Brontosaurus, Written in Stone, When Dinosaurs Ruled,* and *The Last Days of the Dinosaurs,* which won the 2023 AAAS/Subaru Prize for Excellence in Science Writing. A science correspondent for *Smithsonian* and regular contributor to *National Geographic* and *Slate,* Riley is a widely recognized expert on paleontology. She won the 2024 Friend of Darwin Award from the National Center for Science Education.